# Rural Planning in Developing Countries

# Rural Planning in Developing Countries

## Supporting Natural Resource Management and Sustainable Livelihoods

Barry Dalal-Clayton,

David Dent

and Olivier Dubois

iied

# EARTHSCAN

Earthscan Publications Ltd
London • Sterling, VA

First published in the UK and USA in 2003
by Earthscan Publications Ltd

A catalogue record for this book is available from the British Library

ISBN: 1 85383 939 6 paperback
      1 85383 938 8 hardback

Typesetting by MapSet Ltd, Gateshead, UK
Printed and bound in the UK by Creative Print and Design Wales, Ebbw Vale
Cover design by Danny Gillespie

For a full list of publications please contact:

Earthscan Publications Ltd
120 Pentonville Road, London, N1 9JN, UK
Tel: +44 (0)20 7278 0433
Fax: +44 (0)20 7278 1142
Email: earthinfo@earthscan.co.uk
Web: **www.earthscan.co.uk**

22883 Quicksilver Drive, Sterling, VA 20166-2012, USA

Earthscan is an editorially independent subsidiary of Kogan Page Ltd and publishes in
association with WWF-UK and the International Institute for Environment and
Development

Library of Congress Cataloging-in-Publication Data

Dalal-Clayton, D. B. (D. Barry).
    Rural planning in developing countries : supporting natural resource management
  and sustainable livelihoods / Barry Dalal-Clayton, David Dent, and Olivier Dubois.
        p. cm.
    Includes bibliographical references and index.
    ISBN 1-85383-938-8 (hc) — ISBN 1-85383-939-6 (pb)
    1. Rural development—Environmental aspects—Developing countries.
  2. Natural resources—Developing countries—Management. 3. Sustainable
  development—Developing countries. I. Dent, David. II. Dubois, Olivier, 1957-
  III. Title

HN981.C6 D353 2002
307.1'412'091724—dc21

                                                                            2002011680

# Contents

# List of figures, boxes and tables

## FIGURES

## BOXES

# TABLES

# About the authors

**Barry Dalal-Clayton** is director of the Strategies, Planning and Assessment Programme at the International Institute for Environment and Development (IIED) in London where his current work mainly involves research and policy advice on strategies for sustainable development, rural planning and impact assessment. He is also a consultant to a range of development agencies and UN organizations. During the past 30 years, Dr Dalal-Clayton has had broad experience of development challenges, working in many developing countries on land use and rural planning, natural resource management, wildlife management, environmental assessment, soil survey and agriculture. He spent ten years working in Zambia for the UK Overseas Development Administration (ODA, now DFID) and the Norwegian Agency for Development Cooperation (NORAD). He is co-compiler of *Sustainable Development Strategies: A Resource Book* (OECD/UNDP/Earthscan, 2002), lead author, with David Dent, of *Knowledge of the Land* (Oxford University Press, 2001) and author of *Getting to Grips with Green Plans* (Earthscan, 1996) and *Black's Agricultural Dictionary* (A&C Black, 2nd edn 1985).

*International Institute for Environment and Development*
*Endsleigh Street, London, WC1H ODD, UK*
*tel: +44 (0)20 7388 2117; fax: +44 (0)20 7388 2826*
*email: barry.dalal-clayton@iied.org*

**David Dent** is salinity program leader at the Bureau of Rural Sciences in the Department of Agriculture, Fisheries and Forestry – Australia (AFFA). Dr Dent is an international authority on soil survey, land evaluation and land use planning. He has more than 30 years experience as consultant to UN agencies, national governments and NGOs operating in developing countries, and is co-author, with Barry Dalal-Clayton, of *Knowledge of the Land* (Oxford University Press, 2001).

*Bureau of Rural Sciences*
*Department of Agriculture, Fisheries and Forestry – Australia*
*PO Box E11, Kingston, ACT 2604, Australia*
*tel: +61 (0)2 6272 5690; fax: +61 (0)2 6272 5825*
*email: david.dent@brs.gov.au*

**Olivier Dubois** is a forestry officer in the Forestry and Institutions Branch at the Food and Agriculture Organization of the United Nations (FAO) in Rome. He is a specialist in collaborative management of natural resources, with experience in Asia, Africa and Latin America.

*Community Forest Unit*
*UN Food and Agriculture Organization*
*Viale delle Terme di Caracalla, 00100 Rome, Italy*
*tel: +39 (0)6 5705 6494; fax: +39 (0)6 5705 5514*
*email: olivier.dubois@fao.org*

# Preface

In developing countries, rural planning is expected to achieve an eclectic mixture of goals. The more pragmatic ones include effective delivery of public services such as schooling and health care, and the efficient provision of infrastructure for water supply and sanitation, access, communications and electricity. More complex goals include the sustainable use and management of the natural resources that underpin rural livelihoods, and encouragement and support for local enterprises that can broaden the base of the economy, for instance, by provision of facilities and credit. Intangible goals include poverty alleviation, equity and empowerment of local people through participation in the planning process itself. It is a tall order!

To deal with all of these aspects of rural planning in detail within a single volume would be an enormous task and is beyond the scope of this book and our own expertise. Rather, our emphasis is on land resource planning and management of natural resources. We bring together lessons from past and current experience in rural planning from across the developing world. Examples of conventional and more innovative planning practices are taken from Africa, Asia and Latin America.

The book covers the main themes and challenges facing rural planning and local development. In doing so, it does not describe or suggest much that is actually new in terms of rural planning. But it does provide numerous examples of approaches that appear to be working and, collectively, these do represent what can currently be judged as best practice. It also describes approaches that have not been effective. We believe that this study contains much that has never been brought together before, drawing extensively from grey literature, and building on our own work and that of many others.

*Barry Dalal-Clayton*
*David Dent*
*Olivier Dubois*
*March 2002*

# Acknowledgements

This book is the final product of a research project undertaken by the International Institute for Environment and Development (IIED) and funded by the UK Department for International Development (DFID) (see Dalal-Clayton, Dent and Dubois, 2000). The authors are grateful to Peter Roberts (Deputy Chief Engineering Adviser) for his support and to Dr Tanzib Chowdhury (formerly Physical Planning Adviser) for initiating the project.

Contributions were made by several IIED staff: Simon Croxton and Cecilia Tacoli (Sustainable Agriculture and Rural Livelihoods Programme), Dr Camilla Toulmin, Ced Hesse and Thea Hilhorst (Drylands Programme).

We have drawn from published materials, unpublished grey literature and information/inputs from sources and contacts around the world. The book also incorporates lessons from three country case studies undertaken for the project by local teams in Ghana, South Africa and Zimbabwe. Grateful thanks are due to Ian Goldman (Khanya-mrc, Bloemfontein, South Africa), Dr Derek Gunby (Planafric, Bulawayo, Zimbabwe) and Dr George Botchie (ISSER, University of Ghana, Accra) for their inputs and helpful comments.

Chapter 2 draws extensively from a book by Barry Dalal-Clayton and David Dent, *Knowledge of the Land: Land Resources Information and its Use in Rural Development*, Oxford University Press (2001).

# Authors' note

The research project on which this book is based was completed in November 1999 when the original manuscript was prepared. Throughout the text, references are made to materials from a wide range of countries. In particular, the book draws from country case studies undertaken in Ghana, South Africa and Zimbabwe which were completed in mid-1999 and which reflect the position in those countries at that time. We are aware that there have been significant changes in these countries during the last three years in terms of legislation, institutional arrangements and certain key issues (eg land). In preparing this manuscript for publication, wherever possible and to the extent that we have been able to access more recent information, we have updated the text.

# List of acronyms and abbreviations

| | |
|---|---|
| 4Rs | rights, responsibilities, returns/revenues and relationships |
| AGRITEX | Agriculture, Technical and Extension Services (Zimbabwe) |
| ASAL | arid and semi-arid lands |
| BATNA | best alternative to a negotiated agreement |
| CAMPFIRE | Communal Areas Management Programme for Indigenous Resources (Zimbabwe) |
| CBNRM | community-based natural resource management |
| CBO | community-based organization |
| CCC | catchment conservation committee (Kenya) |
| CDF | comprehensive development framework (promoted by the World Bank) |
| CDS | city development strategy |
| CGIAR | Consultative Group on International Agricultural Research |
| CIRAD | Centre for International Cooperation in Agricultural Research for Development (France) |
| CSD | Commission on Sustainable Development (UN) |
| CSIR | Council for Scientific and Industrial Research (Ghana) |
| DA | District Assembly |
| DAC | Development Assistance Committee (OECD) |
| DANIDA | Danish International Development Agency |
| DCC | District Conciliation Court (Burkina Faso) |
| DCE | District Chief Executive (Ghana) |
| DD | democratic decentralization /devolution |
| DDC | District Development Committee |
| DDS | district development strategy |
| DEAP | district environmental action plan |
| DFA | Development Facilitation Act (South Africa) |
| DFID | Department for International Development (UK) |
| DPT | district planning team (Kenya) |
| EIA | environmental impact assessment |
| ESAP | economic structural adjustment programme |
| FAO | Food and Agriculture Organization of the United Nations |
| FAP | flood action plan (Bangladesh) |
| FRA | forest resource accounting |
| FSC | Forest Stewardship Council |
| GDP | gross domestic product |
| GIS | geographical information system |
| GNP | gross national product |

| GT | *gestion de terroir* |
|---|---|
| GTZ | German government agency for international cooperation |
| HDI | Human Development Index |
| HIMA | *hifadhi mazingira* (Swahili phrase meaning 'conserve the environment') programme (Tanzania) |
| HIPC | highly indebted poor country |
| IDA | International Development Association |
| IDZ | intensive development zone |
| IFAD | International Fund for Agricultural Development |
| IIED | International Institute for Environment and Development |
| IISD | International Institute for Sustainable Development |
| ILO | International Labour Organization |
| IRDP | integrated rural development project/programme |
| ITTO | International Tropical Timber Organization |
| IUCN | World Conservation Union |
| KIWASAP | Kifili water and sanitation project (Kenya) |
| KMA | Kumasi Metropolitan Authority (Ghana) |
| LDF | local development fund |
| LDO | land development objective (South Africa) |
| LGU | local government unit |
| LMS | local management structure |
| LRDC | Land Resources Development Centre |
| LTC | Land Tenure Commission (Niger) |
| LTP | Land Treatment Plan (Kenya) |
| MARP | *Methode acceleré de recherche participative* |
| MET | Ministry of Environment and Tourism (Namibia) |
| NCSD | National Council for Sustainable Development |
| NDPC | National Development Planning Commission (Ghana) |
| NFCAP | National Forestry and Conservation Action Plan (Papua New Guinea) |
| NGO | non-governmental organization |
| NRM | natural resource management |
| NSDS | national sustainable development strategy |
| NTFP | non-timber forest product |
| OECD | Organisation for Economic Co-operation and Development |
| PIDA | participatory and integrated development approach |
| PLA | participatory learning and action |
| PRA | participatory rural appraisal |
| PROAFT | Programme for the Protection of Tropical Forests (Mexico) |
| PRSP | Poverty Reduction Strategy Paper |
| RDA | rapid district appraisal (Indonesia) |
| RDC | Rural District Council |
| RDP | rural development project |
| RIDEP | Regional Integrated Development Plan (Tanzania) |
| RM | rural market |
| RRA | rapid rural appraisal |
| RRD | regional rural development (GTZ approach) |

| | |
|---|---|
| SA | stakeholder analysis |
| SARDEP | Sustainable Animal and Range Development Programme (Namibia) |
| SDI | Spatial Development Initiative |
| SFM | sustainable forest management |
| SL | sustainable livelihoods |
| SWC | soil and water conservation |
| SWCB | Soil and Water Conservation Branch (Kenya Ministry of Agriculture, Livestock Development and Marketing) |
| TA | transect area |
| TAA | transect area approach |
| TFAP | tropical forestry action plan |
| TLC | Transitional Local Council (South Africa) |
| TRC | Transitional Rural Council (South Africa) |
| UNCDF | United Nations Capital Development Fund |
| UNCED | United Nations Conference on Environment and Development (known as the Earth Summit, 1992) |
| UNCHS (Habitat) | United Nations Centre for Human Settlements (now UN-HABITAT) |
| UNDP | United Nations Development Programme |
| UNEP | United Nations Environment Programme |
| UNGA | United Nations General Assembly |
| UN-HABITAT | United Nations Human Settlements Programme (formerly UNCHS (Habitat)) |
| UNSO | United Nations Sahelian Office |
| USBR | United States Bureau of Reclamation |
| USDA | United States Department of Agriculture |
| UST | University of Science and Technology (Ghana) |
| VDC | Village Development Committee (Nepal) |
| VIDCO | Village Development Committee (Zimbabwe) |
| WADCO | Ward Development Committee (Zimbabwe) |
| WCMC | World Conservation Monitoring Centre |
| WRRI | Water Resources Research Institute (Ghana) |

# Introduction

This book provides an international perspective on rural planning, focused on developing countries. It examines several conventional fields of development planning and a number of innovative local planning approaches, drawing together lessons from recent and current experience of rural planning and land use in Africa, Asia and Latin America. Opportunities for improving strategic and participatory approaches are presented and the principles of these approaches are underscored. Attention is drawn to the relationships between strategic planning and local economic development, and the ways in which coordinated development planning and management of natural resources can underpin sustainable local livelihoods.

## LESSONS FROM EXPERIENCE

Chapter 1 introduces the different responsibilities assumed by different levels of government – at community, district, provincial and national levels – and the need for a measure of strategic planning, as well as operational activity, at each level. It discusses how concepts of the terms *rural* and *rural planning* vary in different countries, causing confusion between planners, policy-makers and those who have to implement their plans and policies. The state of flux in rural planning is examined: the increasing emphasis on the process rather than producing documents; the evolution of objectives – from encouraging increased production to explicit concerns about equity and the reduction of poverty; the broadening focus – away from a narrow concern with arable cropping, livestock production or forestry, etc, towards the management of natural resources in sustainable production systems; and the weaving of human capital development, infrastructure and social development into integrated rural development strategies.

The experience of experimentation in regional planning since the 1970s is reviewed, including, physical planning approaches, intensive development zones and integrated rural development programmes (IRDPs) (few of which have survived). Following from these, the more recent move to decentralized rural and regional planning is considered. Decentralization is now seen as a key to rural development but the underlying principles are weakly understood and effective institutions are still to be built.

Poverty and livelihoods – both fundamental issues for rural planning – are discussed. The concept of sustainable livelihoods is introduced and, also, the need to identify key stakeholders at the outset of the planning process and to establish their various roles and responsibilities. The importance of secure land tenure and reliable rights of access to the use of resources is also emphasized. The nature and scale of rural–urban linkages is examined: both *spatial links* (flows of people, goods, wastes, money and information) and *sectoral links* (rural activities in urban areas, eg urban agriculture) and the revival of manufacturing and services in rural areas; and the implications for planning for the urban–rural interface are explored.

## CONVENTIONAL, TECHNICAL PLANNING APPROACHES

Chapter 2 reviews the mainstream approaches to resource surveys, land evaluation, impact assessment and land use planning – their advantages and limitations. Potentially, they comprise an array of complementary techniques but, more often than not, they have been undertaken in isolation.

Standard approaches to surveys aim to gather formal information about natural resources: topographic survey; surveys of geology, soil and land use; inventories of forests and biodiversity; and systematic recording of climatic and hydrological data. The main benefits and problems associated with such approaches are discussed, including the adequacy and limitations of the data generated, the way they are presented in technical reports and on maps, their relevance and utility to decision-making. In addition, the data needs of planning at different hierarchical levels are considered.

Approaches to land evaluation are examined, dealing with methods by which basic natural resource survey data are interpreted and combined with other kinds of information, such as market information. The evolution of land evaluation methods is discussed, starting with early approaches driven by the needs of soil conservation, land settlement and irrigation in the US. The main approaches covered include: the land capability classification of the US Soil Conservation Service, the land classification system of the United States Bureau of Reclamation, the Food and Agriculture Organization (FAO) framework for land evaluation, parametric indices (eg the Storie Index Rating), process models, methods of financial and economic evaluation and strategic land evaluation.

Approaches to land use planning include sectoral plans, land allocation procedures, multiple criteria analysis and resource management domains. Examples are provided of the application of both sophisticated and summary land use planning procedures used in village planning in Tanzania, and of the use of a step-by-step planning manual in Bangladesh. The FAO guidelines for land use planning (FAO, 1993) are discussed in some detail together with some proposed redefinitions of the steps involved. There is a also an outline of the array of techniques for impact assessment that have merged in the past 30 years.

A review of decentralized district planning shows how district plans are still frequently sectorally based, lack integration, are undertaken in a top-down manner and, essentially, represent little more than a catalogue of projects for the future. More recent decentralization programmes provide direct funding to district authorities to implement their plans, for instance the new decentralized planning system in Ghana. But there remain formidable problems of institutional capacity to prepare, implement and monitor plans; and a huge investment in training will be needed to equip district staff for their new roles in planning and management.

Planning responses to the challenge of sustainable development include techniques such as sustainability analysis, and the emergence of a range of approaches to developing national and local sustainable development strategies in response to Agenda 21 including comprehensive development frameworks, national visions, provincial and district strategies. Experimentation with sustainable development indicators is also discussed.

The final section examines the pros and cons of the technical approaches described.

## BROADER PARTICIPATION IN PLANNING

Chapter 3 examines the need for and benefits of stakeholder participation in rural planning. 'Top-down' and 'bottom-up' approaches are contrasted. Different perceptions of participation are illustrated through typologies of participation in local development, in policy process and in planning. The use of multi-stakeholder forums to promote participation (for instance, future search conferences) is described. Horizontal participation (eg interactions across interest groups) and vertical participation (eg interaction throughout the hierarchy of decision-making from national to local) are distinguished.

Participatory learning and action (PLA) has blossomed since the 1980s and now encompasses a suite of techniques for diagnostic analysis, planning, implementing and evaluating development activities (eg rapid rural appraisal, participatory rural appraisal, *methode acceleré de recherché participative*). Their principles and the techniques available are described with examples.

Participatory planning – now promoted as an alternative to top-down planning – is discussed in detail. A wide range of examples of different approaches to participatory planning is provided: local level resource planning; and scaling-up and linking bottom-up and top-down planning (eg the use of the 'regional rural development' approach promoted by the German government agency for international cooperation (GTZ), participatory approaches in large-scale projects, the involvement of NGOs as catalysts, the *gestion de terroir* approach in West Africa, participatory planning in Latin America and Landcare in Australia).

The limitations of participation are considered, together with the quality of information arising from participatory approaches, and the costs of participation. The use of stakeholder analysis is discussed as well as the multilateral relations between stakeholders and the issue of power in decision-making.

# A BASIS FOR COLLABORATION

Chapter 4 first reviews the many constraints upon collaboration and opportunities for collaboration in rural development. Before planning begins, there is a need to assess the likely level of competition or collaboration. Conditions for successful negotiation are suggested, together with general principles for interest-based bargaining.

The different ways in which stakeholders view their participation are considered as well as the role of donors as stakeholders. Stakeholder relationships and ways of assessing these are examined. The complex interactions between relationships and power, their influence on negotiation strategies, and some approaches to assessing power and techniques for 'power regulation' are considered.

The impact of many local initiatives to improve collaboration remains limited so long as the institutions that nominally govern rural resources lack the will or capacity to act as effective counterparts. The preconditions to improve this situation include: political will for broad-based interactive participation; renegotiation of roles by stakeholders to accommodate changes; providing an enabling environment with institutions that facilitate rather than dictate the course of rural development; and capacity development (often a contentious issue).

The greatest blocks on broader participation are the disparity in the power of different stakeholders and mutual distrust. So space for dialogue has to be created and sufficient time allowed for confidence building. What this means in operational terms is examined and examples of initiatives to establish dialogue are provided. As one example, we assess the supposed abilities of institutional structures to actually represent local interests and to manage *gestion de terroir* in francophone Africa.

Decisions on the use of natural resources need to be informed by the values placed on their worth by different stakeholders. A range of factors needs to be considered and an array of ways to arrive at a robust and comparable valuation of natural resources is explored. Issues concerning the differentiation of goods and services are discussed: classifying them according to concepts of welfare economics (eg public and private goods, subtractability and excludability); discriminating agriculture from natural resources; the market value of natural capital; and political values, which impact strongly on the way negotiation and collaboration occur in practice – the way such values effect decision-making are examined.

There are ways of combining conventional, financial valuation (which alone is not an adequate basis for rural planning) with a wide range of stakeholders' valuations. These include combining participatory rural appraisal and economic methods, commodity chain analysis, linking resources to users and forest resource accounting.

Next, the institutional framework necessary for participatory planning and management is considered. A pervasive hindrance to rural development is the dispersion of responsibilities across agencies and poor coordination at all levels. Other key issues are incentives (for government officials) to maintain confusion; poor public accountability; inappropriate performance indicators; the way in which donor funding often compounds existing problems; weak managerial and technical capacities and limited financial resources (which increases dependency on donors).

Various concepts and terms concerning decentralization are clarified. The promises of devolution are that it:

- promotes participation, representation and empowerment of marginal groups;
- entails more equitable distribution of benefits and reduces poverty;
- entails more financial autonomy at the local level;
- improves local accountability; and
- increases the effectiveness of local government units in delivering goods and services.

These, and associated realities, in terms of delivery, are explored.

Possible ways to improve the planning, development and management of natural resources in rural areas are set out at three levels of management.

At the *resource/community level*, several important ingredients need to be in place: real power and rights, competence, economic interest and a desire to play a responsible role. Criteria are given by which to assess the probable effectiveness of local institutions involved in local natural resources management, and examples are provided of mechanisms which can complement local elections in enhancing the representation and empowerment of communities.

At the *local government level*, ways are suggested to assess and map institutional capacity and local *institutional topography* (tangible features of the areas under local authorities, which have clear implications for the scope of local development programmes). Initiatives are discussed which help local government to acquire autonomy to undertake development activities and to modify local rules and institutions. Mechanisms to promote greater accountability are suggested; and the need for information and training is considered.

The way that the subsidiarity principle is interpreted in different ways in rural development is examined. Steps to mapping capacity are suggested, and training and information needs are considered.

At the *central government level*, it is argued that the State should let go of power in favour of local initiatives in resource management, facilitating community management. However, it also needs to play an important role in setting out an overall development vision.

## THE WAY FORWARD: CONCLUSIONS AND RECOMMENDATIONS

Chapter 5 summarizes the lessons from half a century of professional natural resources surveys and development planning that are relevant today. They are presented in sections: planning strategy; principles of development planning; natural resource surveys; institutional support; plus some conclusions, including their implications for donors.

These conclusions and recommendations are backed up in the preceding chapters by examples of good practice that can be built upon. They are made with confidence in the case of natural resources information and the principles of land use planning. Inevitably, there is less experience with innovative planning approaches that have emerged only in the last decade.

## Key points

- There is a paramount need for rural planning to operate under a truly domestically driven development vision – not tied to party, ethnic or religious groups; and for coordinated strategies for working toward this vision.
- The sustainable livelihoods concept offers a powerful focus for development planning but it remains to translate the concept into practical guidelines for decision-making and action on the ground.
- Two underlying causes of the general *failure of top-down planning* in poor and emerging countries have been the absence of any local stake or input to the planning process, and the preference of donors to by-pass ineffective local administrations by setting up financially and administratively autonomous project organizations that have further weakened local capacity. In reaction to these failures, *decentralization* and *participation* are now the watchwords.
- *Planning is not a politically neutral, technical activity.* Successful implementation of development plans depends upon common ownership of the problems and the proposed solutions by the people who will be affected. This common ownership may arise from a consensus about the goals and the necessary actions, or from a negotiated compromise between groups with different goals and insights.
- If there is to be negotiation about sustainable development, there must be some forum that commands general respect and legitimacy where all stakeholders can negotiate and contribute to plans. Appropriate *platforms for decision-making* are needed at each level of planning (local, district and national) and all stakeholders must be equipped to participate.

# 1

# Lessons from experience

## RURAL PLANNING: PERSPECTIVES, CONCEPTS AND THE OBJECTIVES AND ROLES OF GOVERNMENT

Different responsibilities are assumed by different levels of government. In recent years, there has been a trend towards shifting increasing responsibility for planning and management from central government agencies to local government but, as well as operational activity, a measure of strategic planning is needed at every level. In outline:

- *Community level:* management of their localities by groups responsible for particular services, eg water point or irrigation committee, school governors. Communities plan and implement activities from their own resources and may contribute to district plans.
- *District level:* representation of the people; delivery of public services and infrastructure projects; management of a substantial district budget; maybe raising local revenue; strategic planning for the district including infrastructure, land use and allocation/regulation of water and other natural resources.
- *Provincial level:* coordination of district plans, financial audit and provision of specialist services not available within districts, eg scientific, engineering and veterinary services.
- *National level:* raising and distribution of revenue for public services; policy-making and strategic planning. In most countries, line ministries remain the main service providers, commonly through staff in provincial outstations.

Rural planning comprises three crucial elements (PlanAfric, 2000):

1 the *content* – the strategies and policies that underlie what rural planning seeks to achieve;
2 the *institutional framework* within which rural planning operates, especially the agencies and people involved and how they interact;
3 the *approach* – often seen in terms of the polarities of a top-down, blueprint approach or a bottom-up approach.

---

**Box 1.1** *Definitions of 'rural' given by service providers in Free State, South Africa*

- The commercial farming area outside urban municipalities, independent of the size of the latter.
- Areas where people still live in villages in traditional systems, eg Thaba 'Nchu and QwaQwa.
- All areas outside of municipal boundaries (therefore mostly including farmland).
- Small towns and their hinterlands where the economic base is mostly agriculture-related.
- Places where people grow what they eat, eg commonages and back yards.
- Areas not having an agricultural context but close enough to services or isolated from large cities.
- The whole Free State.

*Source:* Khanya-mrc (2000)

---

In most developing countries, there has been no integrating review of these aspects of the subject.

Concepts of rural planning vary and this leads to confusion between planners, policy-makers and implementers. This is well illustrated in South Africa where, until 1995, *rural* was defined as all households not living in formally declared towns. Under apartheid, many areas defined as rural were, in reality, urban areas without services. In the era after apartheid, *rural* is now defined as *the sparsely populated areas in which people farm or depend on natural resources, including villages and small towns that are dispersed through these areas.* However, many households fall into both urban and rural categories as they derive their incomes from a range of sources, including migrant labour to towns (Khanya-mrc, 2000). Service providers in Free State provide a range of different definitions of *rural* (Box 1.1) which contribute to the confusion and the lack of a coherent approach to rural planning in this State.

Planning is about preparing for the future. It has to do with setting goals and designing the way to achieve those goals. Everyone needs to plan: individual householders, villages, government authorities (at all levels), entrepreneurs and investors in the private sector, and so forth. Rural planning is concerned with planning for development, land use, the allocation and management of resources, including in the rural–urban interface. In some instances, rural planning equates with regional planning (see the following section); in others it does not.

In Ghana, planning seeks to coordinate across different sectors, identify important inter-relationships and develop collaborative frameworks. Most developing countries have sought to use rural planning as a development tool and, more recently, as a poverty alleviation mechanism. The emphasis now being given to sustainable livelihoods is directly linked to this new thrust within planning. Indeed, in South Africa, planning is required to include a social equity component, in recognition of past inequities.

Now it is frequently argued that the emphasis of planning should be on the process of planning rather than the production of a document (the plan) (eg Klein and Mabin, 1998). This sort of process approach can include goals such as the acquisition of knowledge and perspectives as well as skills (capacity-building) on the part of those involved – and this may influence the course of development as much as any plan-product (Khanya-mrc, 2000).

Planning is often seen as an activity of the state apparatus, attempting to coordinate, rationalize and/or (re)organize human activity and the distribution of resources (Simon, 1990). But this definition *per se* tells us nothing of the motives for such activities – they could be genuinely concerned with promoting development or, at least in part, self-serving in terms of legitimizing existing power relations and state structures.

Certainly, rural planning is in a state of flux. The objectives of planning have evolved over the years from a focus on increased production, through greater efficiency and effectiveness, to explicit concerns about equity and the reduction of poverty and vulnerability. The focus of rural planning has also broadened away from agricultural issues, eg concentrating on water resource allocation and comprehensive watershed management rather than irrigation and drainage. The management of natural resources in sustainable production systems is beginning to replace the independent focus on arable cropping, livestock production or forestry. Human capital development, infrastructure and social development are being woven into integrated rural development strategies.

Peter Roberts (pers. comm.) suggests that it might prove most difficult to win local 'buy-in' on issues of equity since they tend to represent a shift against those who are currently advantaged. In Ghana, the government has addressed this problem (through the Local Government Act No 462, 1993). The powerful are not directly threatened but legal authority is given to District Assemblies (DAs) to promote development of their districts to ensure equity and the reduction of poverty and vulnerability (see Box 2.10). We examine power relations in Chapter 3 (see 'Dealing with power' on page 130) and Chapter 4 (see 'Dealing with relationships and power' on page 139).

## EXPERIENCE OF REGIONAL PLANNING

Since the 1970s, developing countries have seen many initiatives in decentralized rural and regional planning. The reasons behind them have been various but, in most cases, include: concern at the flight of people from rural areas to cities; a desire to reduce regional inequality by some redistribution of resources and by responding to local needs; a wish to secure rural livelihoods by more effective delivery of services like education, health care and agricultural extension; and concern about the degradation of natural resources. These reasons still hold today.

In seeking a way forward for strategic planning at the local level, useful lessons may be learnt from early experiments in regional planning (we take regional to mean a sub-national division of land accommodating physical or social differences – such regions are usually defined by governments as a basis for local administration and planning, eg provinces and districts).

In the early years after independence, the norm for development planning was the five-year plan, and a regional dimension was incorporated through designating regional growth centres coupled with packages of incentives to attract investment. The goal was industrial expansion. Part of the impetus came from the *redistribution with growth* model promoted, in particular by the World Bank and the International Labour Organization (ILO) during the early and mid-1970s. The objective of reducing social inequality was supposed to be met by investment in appropriate sectors as the economy grew, but economies didn't grow, and the master-plan mentality of bureaucracies emphasized physical planning – a geographic zoning of development backed up by regulation (Chenery et al, 1974; Dewar et al, 1986).

Other planning approaches have taken urban centres as their starting point. In the *urban functions in rural development* approach, the strategy for promoting rural development is to develop a network of small, medium-sized and larger centres, each providing centrally located and hierarchically organized services, facilities and infrastructure (Rondinelli and Ruddle, 1978). Rural development was supposed to be stimulated by filling in the missing functions through selective investment in rural towns. Translating this model into practice has been problematic for three main reasons:

1   All urban functions are assumed to benefit the whole surrounding region and all rural households (issues of access, control and economic status are not considered).
2   The methods for selecting key towns have not been clear (they have focused only on the attributes of the towns themselves with no regard for the rural potential).
3   The model is based on generalizations which do not take account of the various roles of urban centres, which are determined by the rural and regional context.

Strategic approaches that have taken rural areas as their starting point include *intensive development zones* (IDZs) promoted, for example, in Zambia in the 1970s with the aim of concentrating resources in a limited number of areas with potential, in the hope that self-sustaining growth would be achieved and the surrounding areas would benefit through multiplier effects. The equivalent in Latin America during this period was *polos de desarrollo* (Development Foci). This attempted to overcome the problems of simple colonization by delimiting large tracts of (usually) marginal land and creating subdivisions to be bought or claimed by interested people. Several plans were implemented under the scheme, but through top-down approaches and with short-term views that frustrated the achievement of the higher development goals while the tracts were occupied and environmental degradation problems ensued. The Spatial Development Initiative (SDI) programme in South Africa (Box 1.2) represents a recent return to such an area-based focus.

---

**Box 1.2** *Spatial Development Initiatives, South Africa*

The SDI programme is a short-term investment strategy that aims to unlock inherent economic potential in specific spatial locations in South Africa. It uses public resources to leverage private sector investment. SDIs are a strategy for boosting investment and kick-starting development in regions of Southern Africa with a high potential for economic growth. The SDI programme consists of ten local SDIs and four intensive development zones (IDZs) at varying stages of delivery. To date, the current portfolio of SDIs has identified 518 investment opportunities valued at R115 billion with the capacity to generate more than 118,000 new jobs. Simply put, the initiatives aim to create jobs and opportunities for real black economic empowerment by encouraging economic growth. They are the practical implementation of the government's economic strategy as set out in its Growth, Employment and Redistribution policy.

*Source:* Khanya-mrc (2000)

---

Such approaches gave way in the 1980s to *integrated rural development programmes* (IRDPs) at the district level (Warren, 1988). Throughout the developing countries, IRDPs grew to be highly dependent on donor assistance and, in particular, came to rely on organizational and procedural autonomy to achieve their objectives. The attraction of by-passing ineffective local administrations is obvious but this very action hindered the development of local institutional capacity to prepare and implement development programmes. In Tanzania for instance, administrative decentralization in the 1970s was accompanied by the preparation of Regional Integrated Development Plans (Box 1.3).

All of these initiatives have been technocratic, top-down exercises and, initially, have been centrally directed. There was a gradual shift away from the national blueprint plan toward provincial or district planning, but this was hamstrung by a chronic lack of capacity at district level. Building this capacity is a long process. None of the early attempts at decentralization successfully incorporated meaningful participation of the supposed beneficiaries in the planning process. In nearly every case, turf wars developed between centrally focused sectoral agencies (eg Department of Agriculture, Department of Transport) and decentralized district authorities. Coordination between the rival agencies has always proved to be difficult (see, eg Sazanami and Newels, 1990, for Pacific countries and Rakodi, 1990, for Central Africa).

A further underlying conceptual problem with much regional development planning has been the assumption that it is the absence of *central places* (or markets) that constrains development. There is increasing recognition that other factors are also important in shaping development – such as ecological capacity, land tenure, crop types and control over crops and access to markets.

Based on an analysis of past experience with regional planning, Simon and Rakodi (1990) identify some of the components needed in strong local planning strategies:

---

**Box 1.3** *Regional planning in Tanzania*

In the 1960s, local administration in Tanzania was provided by poorly funded and ineffective District Councils. In 1972, a new hierarchical administration was instituted with districts amalgamated into 20 regions, each with a political administrator and a Regional Development Committee (comprising representatives and officials) responsible for regional policy formulation. Forty per cent of the national development budget was allocated to the regions. Districts were subordinate but District Development Committees (DDCs) and their Planning Committees were supposed to formulate and implement plans. Regional Integrated Development Plans (RIDEPs) were prepared for each region. Half of these received donor support for implementation but, by 1988, donors had withdrawn their support from all but three (Iringa, Arusha and Kigoma) due to lack of progress. Kleemeier (1988) notes that not one plan was implemented successfully through the decentralized government structure because of the lack of autonomy at the regional level and inadequate resources for management. Nor were the projects able to engender local participation. Local willingness to contribute declined as the government demonstrated its inability to support such projects with staff and money.

The RIDEPs were intended to be strategy documents but they were, in effect, one-off planning exercises. Many were merely shopping lists of projects and they neglected spatial issues. Therefore zonal physical plans were introduced in 1974, to be prepared by the Unit of Regional Planning in the Ministry of Lands, Housing and Urban Development. These were to cover five zones, each comprising two to three regions, and were intended to provide long-term guidance for integrated economic and physical development. However, they had no legal status: did not involve local bodies and, therefore, conflicted with existing proposals; and lacked any administrative framework or resources for implementation.

The establishment of the Rufiji Basin Development Authority in 1975 introduced river basin planning in an attempt to integrate the development of water and land resources in individual catchment areas. Four basins have received some attention but these have been ineffective initiatives, unable to provide real coordination, and have followed the now discredited top-down model of planning.

*Sources:* Kauzeni et al (1993); Kikula et al (1999)

---

- clearly focused objectives and target areas;
- substantial grassroots participation and control over decision-making, as distinct from mere consultation by planners;
- availability of appropriate resources, training and powers;
- orientation towards planning as a process rather than towards the plan as a product;
- on-going monitoring and evaluation of qualitative as well as quantitative dimensions, and in terms of appropriate criteria;
- integration between sectors and ministries to enhance the effectiveness of state activities;
- focus on sustainability as well as increased prosperity;
- replicability without loss of local appropriateness and accountability.

Many of these criteria have been reiterated by subsequent analysts (eg Dalal-Clayton and Dent, 1993; 2001; Bass et al, 1995) and we return to them in subsequent chapters.

---

**Box 1.4** *Integrated rural development programmes*

'The 1970s saw moves to provide more focus for rural development efforts through *integrated rural development projects/programmes* (IRDPs). The concept of the IRDP was to address rural development priorities and needs through a set of mutually supportive components, eg a combination of agronomic packages with credit, development of infrastructure like roads and water supply, commonly with a baseline natural resources survey, sometimes even clarification of land tenure. Many of the ideas now captured within the philosophy of *sustainable development* were embodied within the concept of the IRDP.

Approaches to IRDPs varied as a result of different local conditions and donor philosophies. Many were planned and implemented through an autonomous project body, although initially they may have intended to work with and through local institutions, and many encountered problems of donor dependency.

Without doubt, some IRDPs have had some success in building institutional capacity to undertake planning. For example, under the Serenje, Mpika and Chinsali Districts IRDP in Zambia, "planning, co-ordination and implementation systems were evolved by the district institutions themselves, with the IRDP acting as a catalyst in a flexible *learning-by-doing*, evolutionary approach which was regarded in the country as a model for institutionally sustainable development" (Mellors, 1988).

However, most IRDPs have been integrated in name only. Their component elements have often been little more than a shopping list of essentially independent sub-projects, each of which could have been (and usually were) undertaken separately by responsible government agencies, and simply subsidized by the IRDP's external funds. Whilst IRDP staff responsible for each component may have executed their responsibilities professionally and effectively, management of these diverse activities has proved difficult. IRDP staff have rarely functioned as an integrated team, working on each component collectively so as to bring together their combined skills and engage in an interdisciplinary approach to solve problems and achieve objectives.

In several cases known to us, withdrawal of external funding has led to cessation of activity. No perceptible impact remains in terms of land use patterns or practices. The projects have sunk without leaving a ripple.'

*Source:* Dalal-Clayton and Dent (1993)

---

# A MOVE TO DECENTRALIZED RURAL AND REGIONAL PLANNING

Now there is renewed interest in decentralization. Donors see this as a way of overcoming the deficiencies of the old IRDPs. In particular their propensity to by-pass and, thus, weaken local institutions (Box 1.4).

As Goldman (1998) points out:

*'Decentralisation appeared to offer a locus for integrated rural development, an institution to deal with it (local government), and the potential for downsizing central government and promoting good governance.'*

Ribot (1999) puts forward the concept of *integral local development* to improve the outcomes and sustainability of integrated rural development initiatives. This concept is discussed in Chapter 4 (see 'The local government level' on page 172).

If there is a modern trend towards decentralization, there is certainly a spectrum of quite different situations. Ghana, for instance, completed a national vision for development (the National Development Policy Framework – Ghana Vision 2020) in 1994. There has been an attempt to mirror this vision in policies at all levels and a strongly focused model has been introduced that provides genuine power and places responsibility for rural planning with DAs (see Box 2.10 and Botchie, 2000). In Zimbabwe, there is a mix of devolution to Rural District Councils (RDCs) and deconcentration within central government, but no coordinated strategy for rural development (PlanAfric, 2000). South Africa has a much more centralized approach and is still grappling with the dilemma of the centre versus the provinces and with the lack of viable rural local government institutions that characterized the apartheid years (Khanya-mrc, 2000).

## FOCUS ON POVERTY AND RURAL LIVELIHOODS

Despite the shifts to decentralized planning, policy still focuses on treating the symptoms of poverty rather than addressing the underlying causes. Greater understanding of the underlying causes, addressing inequalities and environmental degradation, and meeting basic needs would seem to be necessary for effective regional development policy. This means empowering local people to manage the resources on which they depend and plan their own development, appropriate decision-making structures and sensible supporting policies. Policy frameworks under which this might be achieved include the *regional rural development* concept developed in the 1980s by the German agency GTZ (German government agency for international cooperation), which is discussed in detail in Chapter 3, and the more recent *livelihoods* focus adopted by the UK Department for International Development (DFID) in its 1997 White Paper on International Development.

### Sustainable livelihoods

DFID's Sustainable Livelihoods group has adopted the following definition:

> '*A livelihood comprises the capabilities, assets (including both material and social resources) and activities required for a means of living. A livelihood is sustainable when it can cope with and recover from stresses and shocks and maintain or enhance its capabilities and assets both now and in the future, while not undermining the natural resource base*' (Carney, 1998).

The sustainable livelihoods framework (Figure 1.1) groups particular components of rural livelihood: their capital assets (Box 1.5), their vulnerability/opportunity context and the institutional structures and processes that may transform livelihoods.

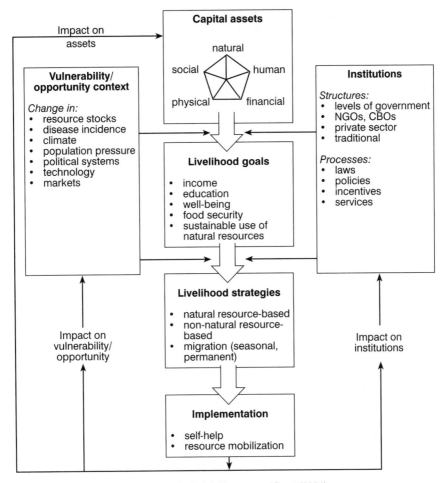

*Source:* Adapted from Carney (1998) by Dalal-Clayton and Dent (2001)

**Figure 1.1** *Sustainable livelihoods framework*

This clustering shows the complexity of the various rural livelihoods and how this complexity might be managed. It also helps to identify opportunities for both external interventions and internal redirection of resources, and assessment of the complementarity of contributions and trade-offs between outcomes. This particular viewpoint appears to call for an area-based rather than sectoral approach to development, and supports the devolution of resources and authority to the district level. Table 1.1 compares the sustainable livelihoods approach with the reality of integrated rural development.

## Stakeholders

How can particular capital assets be maintained, and even increased, without detrimentally affecting others? This question is at the heart of the management

---

**Box 1.5** *Capital assets and the sustainable livelihoods framework*

At the heart of the framework are the capital assets of rural communities:

| | |
|---|---|
| *Natural capital* | the natural resource stocks from which resource flows useful for livelihoods are derived (land, water, wildlife, biodiversity, environmental resources); |
| *Social capital* | the social resources (networks, membership of groups, relationships of trust, access to wider institutions of society) upon which people draw; |
| *Human capital* | the skills, knowledge and health needed to pursue different livelihood strategies; |
| *Physical capital* | the infrastructure (transport, shelter, water, energy and communications) and the production equipment which enable people to pursue their livelihoods; |
| *Financial capital* | moneys (whether savings, supplies of credit or regular remittances or pensions which provide them with different livelihood options. |

These assets are presented schematically (see Figure 1.1), as a five-axis graph (uncalibrated) on which access by different social groups (or even households) to different types of assets can be plotted. Access has a wide interpretation – anything from individual ownership of private goods to customary use rights for groups.

The aim is to promote holistic (rather than sectoral) thinking about the ways in which assets, individually or more often in combination, support sustainable livelihoods. Analysis should reveal much information about the asset status of particular groups and how this is changing over time, eg:

- What changes have occurred over time in the shape and size of the pentagon plot, reflecting changes (improvement or deterioration) in the situation?
- Is there consistency across all axes?
- What changes might occur in the next decade (as population density and the state of resources change)?
- What are the causes of changes and do they vary between different wealth or social groups?

*Source:* Carney (1998)

---

of natural resources and provides a focus for engaging the different stakeholders.[1] They are likely to include special interest groups within rural communities, private operators and both central and local government as primary stakeholders (directly affected by the project) with other stakeholders such as non-governmental organizations (NGOs), research bodies and, sometimes, the state playing more intermediary roles. Experience shows that

---

1 Serageldin and Steer (1994) distinguish between *weak sustainability* – maintaining total capital stocks intact without regard to their composition; *sensible sustainability* – maintaining total stocks intact and avoiding depletion of any capital beyond critical levels; *strong sustainability* – maintaining each component of capital intact (that is any destruction of capital should be replaced); and *absurdly strong sustainability* – no depletion at all of capital stock.

**Table 1.1** *Integrated rural development and sustainable livelihoods compared*

|  | Integrated rural development | Sustainable livelihoods |
|---|---|---|
| Starting point | Structures, areas | People and their existing strengths and constraints |
| Conceptions of poverty | Holistic, multi-dimensional | Multi-dimensional, complex, local. Embraces concepts of risk, power and variability |
| Problem analysis | Undertaken by planning unit in short period of time, viewed as conclusive | Inclusive, iterative process based on holistic livelihood assessment |
| Sectoral scope | Multi-sectoral, single plan, sector involvement established at outset | Small number of entry points, multi-sectoral, many plans, sectoral involvement evolves with project |
| Level of operation | Local, area-based | Both policy and field level with clear links between the two |
| Time taken to prepare projects for donor support | Initial identification rapid; detailed planning time-consuming | Understanding of livelihood options time-consuming. However, projects start as discrete interventions and build on these. Preparation time therefore 'spread' over longer overall project time |
| Time frame | 5–10 years | Longer commitment |
| Coordination | Often donor-driven, dependent on donor funds to implement | Driven by shared objectives and needs identified by those involved |
| Spatial focus | Rural, area-based | Rural areas as part of larger systems |
| Indicators | Production changes, uptake | Production/conservation-oriented, people and outcome-oriented, negotiated |
| Sustainability | Not explicitly considered | Key aspect of livelihoods. Also at political/fiscal levels |
| Environment | Treated as add-on (if at all) | Opportunity to put environment at the heart of livelihood development |
| Capacity-building | Minor concern. Relied on idealized conception of capacity | Major concern |
| Supporting research | Adaptive technical, socio-economic | Livelihood strategy-based action research |

*Source:* Adapted from Carney (1998)

these stakeholders should be identified at the outset of the planning process, and their various roles and goals be established, if they are to work together effectively (see Chapter 4).

**Box 1.6** *Land reform and resettlement programme, Zimbabwe*

Phase 2 of Zimbabwe's Land Reform and Resettlement Programme (LRRP) was launched in September 1998. It aimed to provide a more efficient and rational structure for land allocation through:

- ensuring greater security of tenure to land users;
- promoting investment in land through capital outlays and infrastructure;
- promoting environmentally friendly land utilization;
- retaining a core of efficient large-scale commercial agricultural producers;
- transferring not less than 60 per cent of land from the commercial farming sector to the rest of the population.

It was intended that the LRRP would be set within the National Land Policy Framework (NLPF). The Phase 2 project document was developed through a participatory stakeholder approach. Key contributors were the government, farmers organizations (including the Commercial Farmers' Union), industrial and financial organizations, the Land Task Force of the National Economic Consultative Forum, and civic organizations.

The two-year inception phase for Phase 2 (from October 1998) aimed to acquire farms covering 2.1 million ha for resettlement, and farmer support services were to be provided by government and donors. The white commercial farmers contested the acquisition of most of the farms which had been identified and the donor community failed to deliver on promised financial support. As a result, the resettlement programme slowed down considerably. Using limited resources, the government was only able to acquire 168,264 ha to resettle 4697 families between 1998 and 2000.

The onset of the closely fought general election in March 2000 disrupted the process and major land occupations took place on about a third of the country's 4660 commercial farms. In July 2000, mainly as a political response, the government launched a new 'Fast Track' programme of land redistribution (expected to cover the period July 2000–December 2001) aimed at speeding up the process of land acquisition and taking over 3000 large-scale commercial farms, covering 5 million ha. In effect the Fast Track began a serious challenge to the white domination of the commercial land market. It aimed to:

- identify immediately – for compulsory acquisition – not less than 5 million ha;
- plan and demarcate acquired farms and settle landless peasant households; and
- provide limited basic infrastructure (eg boreholes, dip tanks, roads) and farmer support services (eg tillage, crop packs).

However, many of the technical criteria originally embodied in Phase 2 of the programme fell by the wayside along with some of the legal processes.

Under Fast Track, by June 2001, 112,000 families had been identified to be settled on 3.2 million ha of land, with ZANU party functionaries and supporters in urban areas and (so-called) war veterans being particularly favoured. The genuine 'land hungry' peasantry may not have benefited as much as the propaganda claimed. Where peasants did acquire new land, it lacked basic services (water, roads and social infrastructure, etc) and, in some cases, whole communities left resettlement schemes to return to their original areas.

The Fast Track process brought the total number of people either resettled or identified for resettlement to 127,240 on a total of 5.8 ha since the launch of the first phase of the programme.

*Sources:* Tiisetso Dube and Derek Gunby, PlanAfric, Bulawayo (pers. comm.) and DLARR/DIP (2001)

**Box 1.7** *General effects of the nature of land tenure on the use and conservation of resources*

- If there is open access or if people are using land to which they have no legal or customary rights (either for cultivation as squatters, for grazing or for collection of products), there is no incentive to conserve. Users will take as much as they can get away with.

- In communal tenure systems, individual members of the community have land use rights but these rights are determined by the community as a whole and the land remains the property of the community. Such systems have, in the past, provided sustainable management of diverse resources and spread risk over a wide area of land – particularly useful for the grazing of marginal land. However, communal management is difficult to maintain where there is intense competition for resources and the social sanctions that underpin communal management have been eroded; and there is no incentive for individuals to invest in long-term improvements or perennial crops, especially if the right to use a particular plot is periodically redistributed.

- State land may be managed directly by state agencies or rented/leased under various conditions. Where management is by a bureaucracy, neither this agency nor individuals within it are exposed to the market, nor do they have any incentive to produce or conserve. Concessions to extract products (eg to logging companies) are an incentive to extract as much as possible at least cost and with little or no regard for the consequences. Regulation of this activity by state agencies is rarely effective.

  In many cases, state forests and other reserves were created from land that was previously a communal resource and which is often required as a component of the traditional use of surrounding areas – for example for grazing, hunting, fuel. This creates conflict between the needs of local people and the objectives of the state as interpreted by, eg, the Forest Department.

  Agricultural tenancies may include provisions to protect the land and incentives to produce, eg by the provision of infrastructure (irrigation and drainage works) and inputs (water, agrochemicals). Often the terms of tenancy are a disincentive, especially to invest in long-term management. In general, the shorter-term and less-secure the tenancy, the more the tendency to exploit the land.

- Private ownership, whether allocated under traditional rights or legal deed, gives the user the strongest motive to develop and conserve its resources, at least in principle, and there is a well-established positive correlation between secure land rights and agricultural productivity. However, where farming is not seen as a socially or economically desirable occupation, the incentive to improve and conserve is weak and the short-term profit motive overrides it.

*Source:* John Bruce (pers. comm.)

# LAND TENURE

Land tenure and rights of access to resources are the very foundation of rural livelihoods and, so, must be central to rural planning. In Zimbabwe, for example, the historic legacy of unequal land distribution, rights and planning continues to undermine efforts to plan and develop rural areas. Phase 2 of the Land Reform and Resettlement Programme (LRRP), aims to redress these inequalities (Box 1.6).

The confidence to produce and to invest can only be based on reliable rights to use the land. These rights must be clear, well established, and protected by law. Conservation of resources is promoted by security of use of the same patch of land over the long term. At the same time, a market in land, leasing, or some other customary allocation of use must be flexible enough to accommodate population growth, and give access to land to those willing and able to make use of it (Box 1.7).

Compulsory redistribution of land by the state is one resort but, in recent decades, policy-makers have sought other means for which a broader political consensus is more likely to be achieved. There is an increased emphasis on government-facilitated reallocation of resources through market mechanisms. More significant than private–private shifts in access to resources, is the current trend towards release to individuals and communities of vast resources which the state itself earlier appropriated. The hope is that these resources will be better managed by those who have a direct interest in them. Where these resources are considered by the local community to have been theirs already, the recognition of that ownership may be a better policy than devolution of some title derived from the state.

## Security of tenure

Those who do have access to resources require adequate security of tenure. That is the assurance provided by enforceable rights to invest in the resource and garner the eventual returns on that investment. This applies to both the small farmer and the multinational company, individuals and communities, as well as the state and its agencies.

Security of tenure means:

- *Robust rights:* The resource manager must have rights over the resource that permit genuine control of the resource. Among the most fundamental are the rights to possess the resource; to exclude others where they would interfere with management; and to improve, harvest and fallow the resource. If those who would invest their labour and capital in the resource do not have such rights, as when women in many societies are denied rights to land, their incentives will be undermined.

  For managers, whose economic strategies and opportunities require greater liquidity of assets, and for those who are attractive lending prospects for formal sector lenders, the rights to sell and mortgage land will also be important.

- *Adequate duration:* The manager must be assured of control over the resource until the anticipated benefits of investments have accrued. The length of time required will obviously vary with the type and magnitude of the investment. Many investments take many years to bring returns, with investments in permanent improvements and conservation often paying off only over several generations, and the right of inheritance is therefore often crucial.
- *Legal certainty:* However adequate the rights themselves, they provide little incentive unless there is confidence in enforcement mechanisms and dispute-settlement processes. This applies to both informal mechanisms in local communities and to more formal judicial structures.

To enhance security of tenure, the most important tools are, first, restraint of the state itself and, secondly, tenure reform. National elites have used the asserted ownership of natural resources by the state as a way to grab land from rural people. This is seen in concession regimes that oust customary users from land which is traditionally theirs but, technically, belongs to the state. Certainly the state must retain the right to take land for public purposes but it must be constrained from doing so for private interests and, even in the case of takings for public purposes, without payment of compensation that recognizes the true value of the land. It is increasingly urgent that rural communities demarcate the resources on which they depend, which will often be on land formally belonging to the state, and that the state recognize their rights, individual and communal, in this area.

Where rights are so framed that they fail to provide the necessary assurance, then change may be needed. Tenure reform can proceed in two ways:

1   replacement of the existing tenure by a different model, often a model drawn from more economically developed contexts;
2   adaptation of the existing tenure in specific ways to address perceived inadequacies.

These two approaches are not mutually exclusive in terms of national policy, and different approaches may be applied in different localities.

The replacement approach is direct and, seemingly, final. It is, however, expensive because it generally involves a formalization of land rights, cadastre and establishment of land registries. Usually, individual ownership or leasehold are proposed to replace tenures that are customary and community-based. In practice, few countries have been able to mobilize the resources to effect this and, so, replacement has been a localized strategy in urban and peri-urban areas and special project areas. Where it has taken place, rights of registered individuals have been increased at the expense of their communities, and other individuals' rights in the land have been cut off as well.

It has also proven difficult to erase previous notions of tenurial rights. Often, the result of attempted replacement is the coexistence of inconsistent statutory rights (the replacing regime) and customary tenure regimes (those which were to be replaced). The resultant confusion can undermine security of tenure more dramatically than the failing which the reform attempted to address. On the other hand, there are situations in which existing tenure systems have broken down under intense competition for resources by increasingly diverse claimants, as in peri-urban areas. Then replacement may be the only option, but it should be recognized that there is an almost universal tendency for the powerful to increase their control of resources in the course of this purportedly neutral process.

The adaptation approach is incremental and, generally, relies upon the ability of customary systems to evolve in response to new needs. These largely unwritten systems evolve both through conscious amendment of rules and institutional arrangements by traditional authorities, and through the recognition of changing needs in the settlement of disputes between different users. Both rules and institutions must evolve and, in a market economy, the direction of evolution will be toward fuller property rights for individuals and communities.

Implementation of this approach requires recognition by the state of indigenous custom as the law governing land allocation and use, and empowerment of local institutions to implement this responsibility and to resolve disputes. Adaptation is less costly than replacement but it, too, involves uncertainties: the direction of evolution may be doubtful, and may be vigorously contested within the community. Statutory intervention by the state may still be necessary to address points of confusion; to deal with matters on which national policy is so fundamental as not to tolerate local contradictions, including equality among peoples and genders; and to reform failing institutions.

## Coordinating tenure incentives and disincentives

Twenty years ago, it was not uncommon for tenure to be provided subject to *development conditions* and for the land to be subject to reallocation if those conditions were not met. In retrospect, we can see that these conditions were commonly flawed and, especially, did not allow the flexibility to adjust to market conditions. Monitoring and enforcement were very limited and erratic, and often took place for the wrong reason – someone else wanted the land. The conditions did not create incentives, but undermined them by undermining security of tenure.

## RURAL–URBAN LINKAGES

There is a trend towards the dispersal of manufacturing away from the big cities to green field or small town sites. This offers the prospect of growth in rural incomes and there is a need to understand and take account of such growth patterns and plan accordingly. To promote such growth, the right services and facilities need to be in place (eg infrastructure, credit provision). From the perspective of the livelihoods of vulnerable individuals or households, access to both rural and urban contexts is likely to greatly increase the diversity of opportunity, thus increasing security against the extreme shocks which are common in developing countries. Planning for rural livelihoods and the local economy within the wider regional context requires understanding of the nature and scale of rural–urban linkages. Commentary on such linkages in South Africa is provided in Box 1.8.

---

**Box 1.8** *Rural–urban links in South Africa*

'*Planning health services:* Rural–urban linkages are significant to health planning because out-of-town patients receive most of their services in the towns. As more people leave the farms for the towns, service delivery via mobile clinics becomes very expensive in the farming areas, while the town services become more congested. New solutions for services will have to be found, for example farmers will have to take farm workers to central mobile service points, or farmers' wives who are trained nurses will need to give assistance to the farm workers.

*Planning employment:* Many farm workers leave the farms to live in towns because they hope to get better paid jobs and better health and education services for their families. The planning of local economic development strategies should therefore be closely linked to the farming community to prepare them proactively for the continuous influx of unemployed people, as well as to create opportunities for potential and emerging farmers outside towns. This is not the case at the moment.

*Planning shelter:* 50 years ago, many farmers owned a house in town, to use when they occasionally came to town for church events or to buy and sell goods. This trend has changed with the improvement of vehicles and roads and visits to town have become a normal weekly event. Now many farm workers own a site (even) with a shack or house in town. A senior member of the household, often a grandmother, sometimes lives there with the school-attending children or sometimes the children live alone. Many farmers also encourage their workers to own houses in town because of new legislation that forces farmers to provide permanent accommodation to labour tenants. This has also led to some farm workers commuting on a daily or weekly basis between farms and towns. Housing schemes for farm workers in towns or agricultural villages in the farming areas need to be cooperatively planned by the farming community and the town residents.

*Planning marketing of rural produce:* Large farmer cooperatives played a major role before 1994 in marketing the produce of commercial farmers. After the abolition of control boards and the arrival of the free market system for agriculture, the role of the cooperatives has changed. Cooperatives must now compete with private enterprise to market their produce. Most cooperatives have changed and become input suppliers. Most of the time, the produce of local farmers does not even reach the town. It will go directly via rail or road transport to bigger centres for value-adding or packaging. This has removed a major economic activity from the small rural towns, leaving many people unemployed. However, some innovative commercial farmers have started agro-processing businesses on their farms like dairy processing, mills, sunflower oil processing or abattoirs. These products also by-pass the small towns and go directly to the bigger centres where the markets are, with a detrimental impact on the economy of the town. Local government incentives can lure agro-businesses back to town to stimulate the economy of the town.

*Planning educational facilities:* Before 1994, it was usual practice for white farm children to go to a well-provisioned school in town where they were accommodated in a government-supported hostel. African children, who lived on farms, went to small farm schools with few facilities, equipment and poorly trained teachers. After 1994, all schools were opened to all races and the immediate reaction of the rural whites was to send their children to schools in the bigger centres where there were many whites, or to start local private schools. This again had a major impact on the economy of the small towns.

Instead of going to the nearest town on a Monday or Friday, the trip was now to the big centres where business was also done at the same time. However, there was a major movement of African children to the town schools, in the hope of a securing a better education. Awareness of this migration of scholars is essential to the planning of effective school services.

*Planning transport:* The increase in the numbers of private cars over many years has had a far-reaching impact on rural society. Those who do not own cars find it very difficult to move from the farms to the towns. Taxis do not venture onto the gravel roads between the farms, and buses are rare. There are no trains either. Farm workers are therefore dependent on the farmers for transport to towns.

*Planning linkages:* These linkages highlight the need for the integration of rural and urban services and integrated planning. However, there was hardly any contact between the Transitional Rural Councils and Transitional Local Councils introduced after the end of apartheid – they co-existed with each other, and seemed to operate in different worlds' (Atkinson and Inge, 1997). It remains to be seen if rural and urban services are integrated under the new amalgamated local governments created in December 2000.

*Source:* Khanya-mrc (2000)

For analytical purposes, rural–urban interactions can be divided into two broad categories:

1 *spatial interactions*, between urban and rural areas, including flows of people, goods, money, information and wastes;
2 *sectoral interactions*, including rural activities taking place in urban areas (such as urban agriculture) and activities often classified as urban (such as manufacturing and services) taking place in rural areas.

Two types of interactions, income diversification and migration, are becoming increasingly important in contributing to the livelihood strategies of many groups in low-income countries.

## Income diversification

Farming alone rarely provides a sufficient means of survival in rural areas of low-income countries. Income diversification is increasingly recognized as an enduring and pervasive strategy in developing countries (Ellis, 1998). Non-agricultural rural activities include those carried out on the farm but not related to crop production (such as furniture- and brick-making or brewing, with the products sold in both rural and urban markets) and off-farm activities, all of which have bigger markets around urban centres. Bryceson (1997) defines 'de-agrarianization' as a long-term process involving four main elements: occupational adjustment, income-earning reorientation, social identity transformation and spatial relocation of rural dwellers away from strictly peasant modes of livelihood.

These changes do not necessarily take place simultaneously, or follow similar trajectories. In Brazil, between 1981 and 1990, the average annual growth of non-agricultural rural employment was 6 per cent (compared to 0.7 per cent for agricultural employment), leading to an increase in the number of workers from 3.1 to 5.2 million (Graziano da Silva, 1995). In China, government promotion of 'rural industries' is aimed explicitly at creating employment opportunities in the countryside, reducing migratory pressure on cities. In 1994, industrial production in rural areas was double the output of agriculture, and the number of workers employed in the sector increased from 30 million in 1980 to 123.5 million in 1993 (Yan Zheng, 1995). Rural poverty in Senegal has been linked to a lack of access to non-farm income (Fall and Ba, 1997).

Linkages between non-agricultural and agricultural activities, such as processing of agricultural raw materials and manufacturing of agricultural equipment, tools and inputs are the basis of the most profitable types of non-farm rural employment. This suggests that a rich natural resource base may be as necessary for rural non-agricultural activities, as it is for agriculture (Livingstone, 1997). However, non-farm rural activities are not completely dependent on rural sources and, therefore, are not insulated from pressures at wider levels. For example, the impact of devaluation on the cost of imported inputs and urban supply networks has adversely affected all manner of rural-based activities in Nigeria, from rural transport to grain grinders, from mechanics to photographers. In addition, structural adjustment programmes are forcing small-scale producers into direct competition with exporters, urban consumers and local industry for access to local raw materials (Meagher and Mustapha, 1997).

Opportunities for non-farm production in rural areas usually depend on formal and informal networks. These may be based on income as well as political and/or religious affiliation, ethnicity, household type, gender and generation. In Tanzania, for example, rural women heading their households and widows living alone are often socially marginalized, and may be forced to find employment in unprofitable occupations. Patronage is often a crucial element of access to activities such as intra-regional trade (Seppala, 1996).

## Migration

Internal migration is often seen as essentially rural-to-urban and contributing to uncontrolled growth and related urban management problems in many cities. This preparation has resulted in many policies to control or discourage migration which, generally, have little impact aside from lowering welfare. In fact, most of the growth in urban population is due to natural increase, and rural-to-urban migration is fastest where economic growth is highest, as migrants tend to move to places where they are likely to find employment. For example, secondary urban centres have attracted migrants where they have attracted new investment and industries. In other cases, secondary cities within or close to agricultural regions have benefited from the renewed emphasis on export production (UNCHS, 1996).

The reverse migration from the urban to the rural areas is also, increasingly, frequent and, often driven by, economic decline and increasing poverty. In sub-Saharan Africa, many retrenched urban workers are thought to return to rural home areas, where the cost of living is lower (Potts, 1995) although it is difficult to estimate numbers in the absence of recent census data. Seasonal waged agricultural work in rural areas can also provide employment for low-income urban groups (Kamete, 1998). Temporary and seasonal movement is not reflected in census figures, and this can make static enumerations of rural and urban populations unreliable. Moreover, the complexity in migration direction and duration is matched by variation in the composition of the flows, which reflects wider socio-economic dynamics. The age and gender of who moves and who stays can have a significant impact on source areas in terms of labour availability, remittances, household organization and agricultural production systems.

Increasing movements of people have important implications for understanding the livelihood strategies of the poor, and the changes in the organization and structure of their households. Household membership is usually defined as 'sharing the same pot' under the same roof. However, the strong commitments and obligations between rural-based and urban-based individuals and units show that, in many instances, these are multi-spatial households giving reciprocal support. For example, remittances from urban-based members can be an important income source for the rural-based members who, in turn, may look after the children and property of their migrant relatives. These linkages can be crucial in the livelihood strategies of the poor, but are not usually taken into consideration in policy-making and planning. Absentee land and/or livestock owners are excluded from benefits such as relief or aid measures in case of loss, even in the case of low-income groups for whom such assets are a crucial safety net. Moreover, rural development plans may ignore the fact that migrants can be the key decision-makers while residents may be better described as caretakers with no real power to make decisions over the use and management of local natural resources.

## Implications for planning

Rural–urban linkages are the focus of renewed interest among policy-makers (see Evans, 1990; Gaile, 1992; UNDP/UNCHS, 1995). First, because market-based development strategies, with their emphasis on export-oriented agricultural production, rely on efficient economic linkages connecting producers with external markets. Access to markets is assumed to transform potential demand into effective demand which, in turn, will spur local production. Growing incomes in the agricultural sector will then result in increased demand for services and manufactured goods. From this viewpoint, small towns may play a key role in linking their rural hinterlands with both domestic and international markets, as well as in providing the rural population with non-farm employment opportunities.

A second important reason is the increasing priority given to the decentralization of resources and responsibilities, and to the strengthening of local public institutions. Thus, in addition to their traditional role as infrastructure and service providers, local authorities are also tasked with supporting economic development and poverty alleviation. Poverty alleviation in particular is a tall order since infrastructure provision has been re-focused to that directly related to productive activity, usually at the expense of social infrastructure such as health and education.

The emphasis on market-led development has led economic planners to treat society as an undifferentiated whole. This is mistaken and diverts attention from the most vulnerable groups in both rural and urban areas. For example, spatial proximity to markets does not necessarily improve farmers' access to the inputs and services required to increase agricultural productivity. Access to land, capital and labour may be far more important in determining the extent to which farmers are able to benefit from urban markets. In Paraguay, despite their proximity to the capital city, smallholders' production is hardly stimulated by urban markets. Poor people do not have enough income to invest in cash crops or in production intensification (Zoomers and Kleinpenning, 1996). Patterns of attendance at periodic markets also show that distance is a much less important issue than rural consumers' purchasing power in determining demand for manufactured goods, inputs and services. Some participants in the market are able to enforce mechanisms of control which favour access for specific groups and exclude others. For example, grain markets in South Asia tend to be dominated by powerful local merchants who control access to the means of distribution (transport, sites, capital, credit and information) and, even in the petty retailing subsector, caste and gender are major entry barriers (Harriss-White, 1995).

Migrants returning from urban to rural areas may have acquired new skills. However, their ability to put them into practice and contribute to the development of the rural non-agricultural sector is linked to access to essential assets such as land, capital and labour, and also to social networks which may be crucial in determining access to, and information about, markets.

## The dilemma of planning for the urban–rural interface

The areas surrounding urban communities are, characteristically, part urban and part rural. To varying degrees, they benefit from the infrastructure, facilities and activities of urban areas such as housing areas, industrial sites, manufacturing plants, bus services, shops, etc. At the same time, they also have many rural characteristics with smallholdings and farms growing crops and keeping livestock, areas of woodland or forest, rivers and lakes, and even wildlands.

To date, planning for rural areas has been undertaken in two main ways. First, planning by RDCs (or their equivalents) has usually been undertaken by officials trained in town planning and this has focused almost entirely on the provision of services such as roads, schools, clinics and hospitals. Secondly, planning for the management of rural land and natural resource has been the responsibility of departments of line ministries (agriculture, water affairs, forestry, wildlife, etc). They have used techniques of resource assessment and land use planning, but there has been little coordination between the departments or with District Council planners. Most major rural development projects (some covering extensive areas – even whole districts or provinces) have been funded by donors, and project staff have undertaken their own planning independent of urban or rural authorities.

Peri-urban areas which constitute the urban–rural interface usually come under the control of the urban authorities concerned. Given, their hybrid character, planning for their development also requires a hybrid approach which should draw upon both urban and rural planning experience. In practice, urban and rural planning are completely different schools and there has been little overlap or communication between the two. Clearly, the time has come to bridge the gap.

# 2

# Conventional, technical planning approaches

Rural planners in developing countries – usually based in small towns – have tended to focus on the provision of social infrastructure (roads, schools, clinics, etc). Planning for the rural areas (*sensu stricto*) lying beyond the towns has been mainly a top-down process, usually the domain of government departments concerned with rural development, agriculture and natural resources (eg fisheries, forestry, wildlife, water). Frequently planning has followed procedures, set out in planning manuals, and the focus has largely been on the use of land and natural resources. The current trend towards decentralization is being accompanied by efforts to deconcentrate and devolve planning functions and more attention is being given to the effectiveness of delivery and the historical legacies that remain.

Plans in the past have usually been made in the offices of government departments, remote from the areas being planned or the people who would be affected and, usually, without their involvement.

The planning process has been technical, relying first on the gathering of basic information about natural resources and socio-economic conditions in the areas concerned – followed by analysis and interpretation (eg land evaluation – see the section on this topic on pages 35–44). The information has been provided by natural resources professionals and institutions. The methods of survey and interpretation are well tried and reliable, if carried out to a good professional standard, but there is often remarkably little evidence of the actual use of their information in the subsequent planning.

In this chapter we review the main methods of survey and interpretation of natural resources data; consider the main planning approaches; and, finally, discuss the pros and cons of the conventional approaches. We recognize that good quality information is also needed on a wide range of other subjects in rural planning, but the approaches for this are beyond the scope of this chapter. The sustainable livelihoods framework (Figure 1.1) might ensure a comprehensive and balanced approach to information needs. Its use in planning would help to demonstrate the need for good data linked to all the main capital assets areas (natural, human, financial, physical and social) and would, therefore include such matters as demographic information, local skills availability, rural–urban linkages, etc. Natural resources information is but one vital element among the five assets and it is necessary to make connections and inter-linkages.

## RESOURCE SURVEYS FOR PLANNING

Formal information on natural resources is gathered by surveys: topographic survey, surveys of geology, soils and land use, forest inventories, biological diversity stocktakings and systematic recording of climatic and hydrological data. These data are presented in detailed technical reports and as maps. It is a characteristic of developing countries that data about natural resources and their use are incomplete and unreliable, and much that was gathered has now been lost (Box 2.1).

The basic data are rarely in a form that can be used directly by decision-makers, but interpretation of these data in terms of development options and their consequences has rarely been provided. As a result, decisions have been taken, and are still being taken, in ignorance.

The dilemma has been recognized for some time. More than 30 years ago, Robertson and Stoner (1970) commented:

> *'It is alarming to observe how little of the land resource data investigated and mapped is actually used in development plans – though one must recognise that its lack of use is by no means always related to irrelevance to the objectives in mind or to the form in which it is presented...*
>
> *... frequently, public organisations gather routine data without any clear idea whether it can be used, until the point is reached where it cannot possibly be handled. And yet the collecting process continues. Experience teaches us to beware of massive data gathering programmes. Such programmes can be expansionist and obsessive in their demands on the investigators' time until merely collecting data becomes an end in itself.'*

In seeking to establish how this situation has arisen and how it may be rectified, Dalal-Clayton and Dent (2001) address both sides of the question: first, the methods of survey, presentation and targeting of natural resources information and, secondly, the needs and capabilities of decision-makers within the institutional framework of land use planning.

Often, large surveys have been carried out without clearly establishing who will be using the information, how it will be used, and what is the capacity of institutions and individuals to make use of it. Under these circumstances, it is hardly surprising that little or no use is made of the findings. Clearly, at the outset of a survey programme, there is a need to establish what data are needed, by whom and for what purposes. Box 2.2 provides a checklist of specifications for natural resources surveys to avoid the pitfalls. A workshop held in 1997 for district development planners in Ghana revealed that they had clear ideas on the kind of information they required for more effective planning, where some of it might be available and what additional information was required (Table 2.1).

**Box 2.1** *Atrophy of natural resource survey information*

Most of the formal information about land resources in developing countries was gathered by expatriates: during the golden age of natural resources surveys (1956–69) as members of large interdisciplinary teams mounted by agencies of the colonial or former colonial power (eg the UK-based Land Resources Development Centre (LRDC)); later seconded to local survey departments of independent countries under aid agency support programmes; or operating for consultancy companies.

Much of the data acquired was archived at the time within the offices of the survey organizations concerned (in-country in the case of local survey organizations and overseas in the case of consulting companies and international survey organizations). The expatriate staff had laboured hard to acquire the data, had a vested interest in them and understood their value and utility for planning. The data were summarized in the final reports presented to the clients (usually the government) in the host country, but there was usually no in-country institutional ownership of these data. Even where in-country databases were set up (quite common), the host country seldom made any provision for their maintenance – so they have rarely survived. The former colonial survey institutions have been mainly disbanded and few people maintain any interest in the archived data. As a result, the data they once held are now scattered. What remains is mostly in the hands of individuals and a few archives but not as an organized body of data. In the case of data gathered by LRDC, some of these have recently been placed in DFID public records.

During the last 20 years, most natural resource survey operations funded by donors required that host country professionals were appointed as counterparts to expatriate members of the survey teams, usually in a junior capacity – seldom in a leading role. The counterparts usually received on-the-job training and commonly were sponsored for post-graduate training overseas. The goal of such training was to provide a skills and capacity base in the countries concerned. But few of these professionals are still practising. They are either coming to the end of their careers, are retired, have passed away or have moved into other areas of employment. In any case, few have a strong interest in maintaining the data gathered in the past (whether during the colonial period or since independence) by expatriates. Nowadays, most natural resource survey institutes in developing countries are unable to function effectively (skilled staff and financial resources are mainly lacking and equipment is defunct); donors no longer support such survey operations and archive maintenance has a low priority.

There has also been an underlying institutional problem with natural resources information. It has never been part of the decision-making process. Where it was gathered for a job-in-hand, it might have been used immediately, then shelved. Often it was gathered as background to future development that never happened. So the real potential was understood only by the technical data gatherers and they were rarely part of the decision-making team. Others, particularly senior decision-makers and politicians, had no real appreciation of its value and gave no attention to ensuring the maintenance and survival of data collections.

Privatization and disbanding of organizations like Huntings Surveys, mean that it will be difficult to resurrect the data. We have, at most, five years to do it.

In rural situations, there are many different decision-makers and they have different needs. Different kinds of information in different degrees of detail are also needed at different stages in rural planning and in development projects. Ideally, decision-makers need to use the fund of information iteratively, so there

**Table 2.1** *Natural resources information needs identified by district planners in Ghana*

| Resource | Information needed | Possible information sources |
|---|---|---|
| Land | Land ownership, housing stock; land use, land suitability, land potential, soil quality, natural resources inventory; population | Land Evaluation Board, Land Title, District Revenue Records, Surveys Department, Town and Country Planning, Soils Research Institute, Universities, Ministry of Agriculture, Lands Commission |
| Water | Sources of pollution, watershed information, dispersal of waste, extent of pollution, water-borne diseases | Environmental Protection Agency, KMA Waste Department, UST Civil Engineering, Clinics, Chiefs, District Authorities, CSIR-WRRI |
| Timber | Species, coverage, timber concessions, rate of exploitation | Forestry Research Institute, FIBP, Forestry Department, Ghana Timber Board |
| Non-timber forest goods | Suitable species, inventory, rate of exploitation | Herbalists Association, Department of Game and Wildlife, Forestry Department, Institute of Science and Technology |
| Fish | Species, resources, practices | Fisheries Department, Universities |
| Sand and stone | Winning sites, contractors, demand, impact | Mineral Commission, Mines Department, Environmental Protection Agency, Town and Country Planning, Sand & Stone Association |
| Refuse and waste disposal | Housing, access roads, type and volume, disposal sites, collection sites | KMA, District Environmental Health Officers |

*Source:* NRI/UST (1997)

is a need for interactive ways of presenting and manipulating the data to match them with the questions of the moment.

When it comes to information about natural resources, three situations are common:

1   there are no data;
2   relevant data exist but the people making the decisions don't know about them or don't have access to them;
3   data exist, are accessible, but are not comprehensible to the people who make policy and land use decisions. Moreover, data are of variable quality but the decision-makers and, even, professionals in allied fields, have no way of knowing which are reliable and which are not.

With the expansion of aid agency programmes in the 1970s, organizations like the Land Resources Division of the British Directorate of Overseas Surveys mounted substantial professional teams to undertake aerial photography,

---

**Box 2.2** *Checklist: specifications for natural resources surveys*

1   **What is the information going to be used for?**
    - Are the goals of the development programme and the specific uses of the natural resources data clearly defined?

2   **Who will use the data?**
    - There may be several categories of users, each with different requirements. Have these users been involved in defining the goals and specifying the data needed, the scale and precision of survey and format of the data?
    - Are these people able to interpret the data requested and check their quality?

3   **When will the data be used?**
    - How soon are they needed?

4   **Do relevant data already exist?**
    - Where are they held, are they accessible? Do they cover the whole area of interest? Are they up to date, reliable and at an appropriate scale – and who has checked this?

5   **What useful information is held within the local community?**
    - What information is already used in land use decision-making?
    - Are local methods of data collection, classification and analysis appropriate?
    - Are the proposed professional staff experienced in techniques of participatory inquiry that can elicit locally held information?
    - On completion of the survey, the results should be explained to all interested parties, not least to local people, both for their information and for immediate correction of errors that are obvious with the benefit of local experience.

6   **Who is best placed and best qualified to provide the data needed?**
    - Are there local institutions, NGOs, local consultants or national specialist organizations that can undertake or contribute to the work?
    - If expatriate consultants are selected, can they provide training for in-country staff?
    - Can they supply data of the quality needed in time and within budget?

7   **Are the proposed methods of survey the most appropriate?**
    - What alternatives have been considered? Will the data be compatible with existing key data?

8   **Are there adequate resources to complete the work to the standard needed and to implement any recommendations?**

9   **What support is there for the project at every relevant level of government and across the local community?**

*Source:* Dalal-Clayton and Dent (1993)

---

topographic and natural resources surveys, mainly, but not exclusively, in Commonwealth countries. Published works include land systems atlases of Lesotho (Bawden and Carrol, 1968), Swaziland (Murdoch et al, 1971), part of Kenya (Scott et al, 1971), Uganda (Ollier et al, 1969) and Indonesia (Land Resources Department – Bina Programme, 1990). Other land resources studies included Ethiopia (Makin et al, 1975), north east Nigeria (Bawden et al, 1972), western states of Nigeria (Murdoch et al, 1976), central Nigeria (Hill, 1978/79) and Zambia (Mansfield et al, 1975/76). Land resource inventories carried out by consultant companies include those by Huntings in Sri Lanka (Hunting Survey

Corporation, 1962) and Sudan (Hunting Technical Services, 1974; 1976; 1977) funded by multilateral aid and in Nepal (Kenting Earth Sciences, 1986) under a bilateral agreement. Specialist agencies of the United Nations, especially FAO, also undertook substantial programmes of survey.

Little development seemed to result from these resource inventories but an exception is the Mahaweli power and irrigation project in Sri Lanka that can be traced back to the surveys carried out by the Hunting Survey Corporation under the Colombo Plan.[1] Development projects were more often directly associated with surveys for irrigation – which universally adopted a conventional hydrological and engineering appraisal, soil survey and financial appraisal according to the procedure of the United States Bureau of Reclamation (1953).

As well as the regular programmes of survey organizations in developing countries, there was an exponential growth in the 1970s in development projects such as integrated rural development programmes funded by aid agencies. Following the debacle of the Tanganyika Groundwater Scheme of the late 1940s and 1950s, stemming in part from ignorance of soils, donors usually required that these projects be preceded by resource assessments as a basis for planning and implementation. These surveys were undertaken mostly by consultancy companies or external organizations, the survey methods used and the reporting procedures usually prescribed by the aid agency or, sometimes, the methods were recommended by the consultants. As far as soil surveys are concerned, they have been based on established international systems of classification and land evaluation put forward by the same small group of professionals. There is little evidence of consultation with the host governments about the methods and products (some would argue that such consultation would not have helped much!) or prior consultation in-country with potential users of the information. As far as planning aspects are concerned, there is almost no indication of the involvement of project beneficiaries.

Developments since the early 1970s include the general use of remote sensing for a range of surveys, and procedures for interpretation of the basic data beyond the mechanical application of land capability classification or the United States Bureau of Reclamation (USBR) payment capacity criterion. The FAO Framework for Land Evaluation (FAO, 1976) recognized that land cannot be graded from best to worst regardless of the purpose, and procedures were developed for the characterization of land use types and land evaluation for these specified land use types (FAO, 1984a; 1985; 1991). The FAO procedure, in its turn, has often been applied ritually to survey data with no perceptible increase in the utility of the information to farmers or policy-makers. Two notable innovations from the UK-based Land Resources Development Centre were the presentation of information from the huge land systems survey of central Nigeria (Hill, 1978/79) in terms of crop options and agricultural development possibilities; and the focusing of data from the land resources study of Tabora Region, Tanzania, through algorithms calculating human- and

---

1 The Colombo Plan: an arrangement to provide technical assistance and finance for development projects in South and Southeast Asia, established in Ceylon in 1951.

stock-carrying capacities, to identify specific areas which were over-populated or which had quantified opportunities for resettlement (Corker, 1982).

Development has turned out to be not so simple as once was thought. Some of the goals of development now seem illusory; the constraints more and more intractable; and the contribution of natural resources information disappointing in the absence of ways and means of using it. This is no better illustrated than in the former Soviet Union where a superb database of surveys at a scale of 1:50,000 for individual collective farms and 1:100,000 for whole administrative units was tied to a powerful system of central planning. Yet this failed to prevent the desertification of the Aral Sea and other environmental disasters against which the failure of the Tanganyika Groundnut Scheme in the 1940s pales to insignificance.

Development failures have been much more often due to misuse and misinterpretation of survey data than failures of the data themselves. Commonly, detailed development work has been undertaken on the basis of information gathered at reconnaissance scale. In other cases, basic data have remained unused because they did not match the needs of the decision-makers. Hills (1981) drew attention to the weakness of natural resources data that cannot be applied in the economic models that planners and policy-makers use. Since then, quantitative land evaluation and crop modelling have come of age (FAO, 1978; Beek et al, 1987; Kassam et al, 1991; Driessen and Konijn, 1991). The obvious advantage of being able to make quantitative predictions about crop production is that these data can be fed into benefit–cost models, eg Querner and Feddes (1989). However, current simulation models demand a lot of quantitative data that are available only from instrumented sites, which are rare, and the methods are still the province of a few specialists and their computers.

A hierarchy of data is required for planning at different levels: generalized at national level, increasingly detailed for application at district, catchment and field level. Strides have been made in computer-based information storage and management, though bedevilled by lack of practical approaches to realistic ownership, archiving and access.

The apparently insuperable problem of providing all the necessary detail at the catchment (watershed) and field scale (there are not enough natural resource specialists to go around) may be solved by do-it-yourself survey kits developed by professionals for use in the local community, backed up by expert systems that enable local people to interpret their data for themselves.

## LAND EVALUATION

Over the last 30 years, there has been a shift of emphasis in the work of natural resources professionals and institutions. Initially they concentrated on the collection of basic, specialist data: topographic surveys produced topographic maps, geological surveys produced geological maps, soil surveys produced soil maps, and their responsibilities seemed to end there. In other words, they saw 'the land, what it is'. But these basic data do not answer the next, more complex and difficult question faced by policy-makers, land use planners and managers:

'What is it good for?' Indeed there are sequences of questions. Planners and policy-makers at national level might ask:

- Where are the useful areas and what is their extent?
- For which uses are they suitable and what is their potential?
- Is there land enough to meet present and future needs for food, industrial raw materials, timber, water supply and urban and industrial development?
- How much of the suitable land will be needed to meet each specified demand and how much can be diverted to other uses?

At farm level, the land user needs to know:

- What are the opportunities to increase production and what are the limitations?
- Where can the best returns from increased input be obtained?
- What investment is needed to obtain these returns in terms of capital, equipment, labour and management?
- What risks are associated with each option (eg from weather, land degradation, pests and diseases) and what measures are needed to manage the risks?

To answer these questions, a range of basic land resources data has to be interpreted and combined with other kinds of information, such as market information. Expert knowledge of land response to management can be used to interpret basic data, providing information that is, at once, more accessible and more focused on the problem in hand. Four approaches have been widely adopted: Land Capability Classification, the FAO Framework for Land Evaluation, the USBR land classification system for irrigation schemes, and various parametric indices. More recently, decision trees have been advocated as a more transparent way of using expert knowledge so that it can be built upon by the manager or decision-maker in the field.

Physical land evaluation tries to explain and predict the potential of land for one or more uses by systematic comparison of the requirements of land use with the qualities of the land. The end product is an index of potential performance in terms of *capability* to support broadly defined categories of use, *suitability* for some specified land use, or *productivity* (eg crop yield) of a specified land use. In this way, the range of feasible land use options may be identified. Where economic appraisal has been demanded, this has been tacked on without much change in procedure, either from natural resources specialists or economists.

## Land capability classification

The best known and most widely used method of land evaluation is Land Capability Classification, originally developed by the United States Soil Conservation Service in the 1930s to interpret soil maps for farm planning (Hockensmith and Steele, 1949). It has been adopted and, sometimes, modified by survey organizations in many developing countries. The definitive account is given by Klingebiel and Montgomery (1961).

There are three levels of classification. At the highest level, land is classified according to the *degree* of its limitations for sustained use and the soil conservation measures necessary to maintain it in productive use (Table 2.2). The limitations to using land that are considered are those which it is not feasible for the farmer to correct (eg climate, slope, soil depth, liability to flooding) and these are recognized individually at the next, subclass, level. A third level of classification, within the subclass, is the capability unit which groups *soils that require similar management* and are suitable for similar crops.

Land allocated to a particular class has capability for the land use defined for that class (eg Class I: arable, no restrictions) and all uses allowed for lower classes, so Class I is also suitable for grazing, forestry and wildlife. The defining land use for the class does not necessarily indicate which use is the most productive or profitable; for example, the world's most sought-after coffee is grown only on Blue Mountain in Jamaica, on land that would be classified as Class VII!

Decision-makers of all kinds are much more comfortable with land capability maps than with the foundation land resource data. In an article called 'We don't want soil maps. Just give us land capability', Woode (1981) wrote:

> '*It is probably true to say that in Zambia hardly anybody reads soil survey reports except the Senior Soil Surveyor and a few visiting consultants. Soil maps are rarely unfolded and the various soil names are known only to the surveyors who described them. But the land capability maps are used constantly...*'

After all, class I is obviously the best; class III is less good; and class VI is rubbish, isn't it? Alas, no, but the assumptions of the system are hidden and the shortcomings are not immediately obvious. Perhaps its most obvious shortcoming is that land cannot be graded from best to worst, irrespective of the kind of management. Some kinds of use have special requirements and tolerances that others do not have, for example:

- Rice enjoys prolonged flooding; other cereals will not tolerate waterlogging during their period of active growth.
- Tea, sugar cane and oil palm need efficient transport to processing plants and so have a minimum area requirement; grain grown for subsistence does not.
- For mechanized operations, stones and rock outcrops are limiting; but with oxen or hand-hoeing the farmer can work round them.

The arable bias of Land Capability Classification and the very generalized information presented do not help choice between alternative uses, except to eliminate the grossly unsuitable. Land use, productivity and profitability are often poorly correlated with land capability class. Indeed, no one-shot land evaluation can provide the information needed to choose between several land use options and, thus, match land use closely with land suitability. These

**Table 2.2** *Land capability classes defined by the USDA Soil Conservation Service*

| Class | Description |
|---|---|
| I | Soils with few limitations |
| II | Soils with limitations that reduce the choice of crops or require simple soil conservation practices |
| III | Soils with severe limitations that reduce the choice of crops and/or require special conservation practices |
| IV | Soils with very severe limitations that restrict the choice of crops and/or require very careful management |
| V | Soils with little or no erosion hazard but with other limitations that limit their use largely to pasture, range, woodland or wildlife |
| VI | Soils with very severe limitations that restrict their use to pasture, range, woodland or wildlife |
| VII | Soils with very severe limitations that restrict their use to range, woodland or wildlife |
| VIII | Soils and landforms with limitations that preclude commercial crops and restrict their use to recreation, wildlife and water supply |

There is an in-built assumption that the most desirable land use is arable cropping requiring no special soil conservation practices. This determines the choice of limiting factors and the values of limiting land characteristics assigned as class boundaries:

| Class | | |
|---|---|---|
| I | Arable (all crops, no conservation practices) | Most desirable |
| II, III, IV | Arable (increasingly costly conservation practices and/or restricted choice of crops) | |
| VI | Improved pastures | |
| VII | Grazing of natural range, or forestry | |
| VIII | Recreation, wildlife, water catchment | Least desirable |

Class V is an oddball for limitations other than erosion. In practice, it is used for wetland.

*Source:* Klingebiel and Montgomery (1961)

objections were addressed in the FAO Framework for land evaluation (see 'FAO framework for land evaluation', page 39).

## The USBR system

The land classification system of the Bureau of Reclamation of the US Department of the Interior (USBR, 1953) was developed for planning irrigation projects. It classifies land in terms of its *payment capacity* – the money remaining for the farmer after all costs except water charges are met and after making an allowance for family living costs. This was an early attempt to integrate physical and financial criteria of land suitability and remained the standard method of evaluation for irrigation projects for more than 30 years.

Classes 1 to 3 have progressively lower positive payment capacities; class 4 designates restricted land use or special engineering needs; class 5 is a holding class pending further investigation; and class 6 is not suitable for irrigation, as it doesn't pay. Classification proceeds directly by survey of the relevant land characteristics and USBR prescribes the scale, accuracy and survey intensity for different purposes. The USBR system works only for a single use within a specific scheme. The parameters vary from scheme to scheme and payment capacity also depends on the costs and farm-gate prices at the time of assessment.

## FAO framework for land evaluation

The first principle of the framework (FAO, 1976) is that evaluation is for a specified *land use type* – a system of management relevant to local conditions in terms of the physical environment and social acceptability – so the first step is to identify and define promising land use types and establish their land requirements. For land use planning, there is also a need to know requirements for labour, capital and infrastructure – so the definition of land use types becomes a substantial, interdisciplinary task.

Knowing the land requirements, relevant information about the land is assembled and the various land suitability classes are arrived at by matching the requirements of the land use type with the qualities of each land mapping unit. The situation becomes more complicated where a land use type depends on several contrasting kinds of land, eg extensive grazing systems may require separate land areas to provide forage in the wet season and in the dry season. The structure of the framework is outlined in Box 2.3.

Early applications of the framework were qualitative and matched land use requirements with land qualities in the same way as in land capability classification. Typically, one or more diagnostic land characteristics (eg soil drainage class, slope angle) have been used as surrogates for land qualities, and limiting values for each land quality are fixed by expert judgement combined with field calibration. The principal judgement is to determine the cut-off point between suitability and unsuitability for each attribute. The next step is to determine the point at which the attribute changes from having no effect, to having a significant effect on production. A refinement is to begin with a cropping calendar for the land use type and, from this, establish critical periods of the year for different qualities, eg trafficability at sowing and harvest, a dry spell for ripening and so forth.

The Law of the Minimum works remarkably well but many practitioners prefer a matching procedure that allows some compensation between different land qualities. Various procedures for rating limitations have been developed (eg Sys et al, 1991).

Detailed procedures for establishing land suitability are given in the series of FAO Guidelines for Land Evaluation (for rainfed agriculture 1984a, forestry 1984b, irrigated agriculture 1985, and extensive grazing 1991) and by Sys et al (1991, 1993). The end-product is reassuringly familiar: a map of land suitability looks like a simplified soil map but, in the absence of short cuts that can be taken

---

**Box 2.3** *Structure of the FAO framework for land evaluation*

The target set by the FAO framework is a four category evaluation:

**Land suitability orders.** The first ordering is into SUITABLE or NOT SUITABLE for a specified land use type.

*Suitable* means that sustained use of the kind under consideration will yield benefits which justify the inputs without risk of unacceptable damage to land resources.

*Not suitable* means that the kind of land use is impracticable, or would cause unacceptable degradation of land resources, or that the value of expected benefits does not justify the expected costs of needed inputs.

**Land suitability classes.** These reflect degrees of suitability. Experience of testing land suitability evaluations against crop performance does not support detailed subdivision. Within the suitable order, FAO recommends not more than three classes:

*Class S1*   Land having no significant limitations to sustained use, or with only minor limitations that will not significantly reduce productivity or benefits and will not raise inputs above an acceptable level.
*Class S2*   Land having limitations that, in aggregate, will reduce productivity or benefits and will increase required inputs so that the advantage to be gained from the land use, though still attractive, will be less than that expected on Class S1 land.
*Class S3*   Land having limitations that, in aggregate, are so severe that expenditure on the land use will be only marginally justified.

Within the Not Suitable order, there are two classes:

*Class N1*   Currently not suitable. This land could be used for the purpose under consideration but the social or economic cost is, at present, unjustified.
*Class N2*   Permanently not suitable. Land having limitations that appear so severe that sustained use is not possible.

**Land suitability subclasses.** These reflect the kinds of limitations, such as water deficiency, erosion hazard, eg S2w, S3e. There are no subclasses within S1.

**Land suitability units.** These are subdivisions of a subclass which differ in their response to management and so are significant at the farm level. They are distinguished by Arabic numbers, eg S2e – 1, S2e – 2.

*Source:* FAO (1976)

---

only by old hands, the procedures are time-consuming and, sometimes, opaque. The choice of factors may be arbitrary, the factors poorly defined and, where weightings are applied, the precision of any resulting numbers is spurious: the methods are essentially qualitative. And after all this, knowledge that a parcel of land is, say, suitability class S2 for sorghum but class S3 for cotton, is not a lot more useful to a farmer or a policy-maker than news that its land capability class is III. Various workers have attempted to develop more rigorous ways of weighting and judging land qualities and determining land suitability, eg through the use of *fuzzy sets* (Chang and Burrough, 1987; Triantafilis and McBratney, 1993).

The clarity of thought behind the FAO Framework has had a profound influence on land evaluation, paving the way for simulation models (see 'Process models', page 42). However, modelling of all the land qualities relevant for land evaluation requires detailed spatial and temporal data for many individual land characteristics, and these are not often available.

To command confidence, land evaluations are usually calibrated against some measure of land performance. In the case of agricultural land, crop yields provide the obvious yardstick. Land use, productivity and profitability are often poorly correlated with land capability or land suitability class (Aitken, 1983; Burnham et al, 1987). Rather than proceed through the intermediate stage of capability/suitability classification, *land potential ratings* arrive at local comparative ratings of land mapping units by direct calculation of an index of performance or yield against a locally established standard (the best lands in the area, under good management) modified by the costs of measures to overcome or minimize the effects of land limitations and the costs resulting from continuing limitation (Soil Survey Staff, 1951; McCormack and Stocking, 1986; Stocking and McCormack, 1986).

## Parametric indices

Parametric methods of land suitability assessment assume that land suitability or performance is determined by only a few *significant* factors. The effect of each individual factor is expressed as a response function. A host of parametric systems has been developed. In the best known, the Storie Index Rating, a single numerical rating is arrived at by multiplying factors representing the *character of the soil profile, topsoil texture, slope* and a fourth factor for any other significant impediments such as salinity or erosion:

$$SIR = A \times B \times C \times X$$

Each factor is scored as a percentage but multiplied as a decimal. The time factor is expressed as a percentage. The key factors, their rating and weighting are defined by the expert to produce an outcome in accord with experience and acceptable to users (Storie, 1933; 1978).

Subjective decisions are taken by the expert at several stages: in the selection of properties to be used, the valuation of each factor, the formulation of the equation and, not least, the translation of the final numerical value into operational terms. The Storie Index illustrates common shortcomings of parametric indices. It uses a compound factor, *character of soil profile*, which includes factors that are not independent variables and are used more than once in the calculation. Above all, the functions are developed and tested for one application, in one area and at one time. They do not travel well.

From the point of view of the land use planner, the system is easy for a non-specialist to apply. But the final numbers appear as if by magic; the assumptions are hidden; and the logic is difficult to retrace. If performance really is dependent on just a few characteristics, it is better to say so explicitly.

Parametric indices represent an evolutionary dead end in land evaluation, although they represent the beginnings of calculation. The way forward lies, on the one hand, in more transparent knowledge-based systems and, on the other hand, in quantitative modelling of physiological processes.

## Process models

Process models (or simulation models) have been developed to predict, for example, crop production, risk, or inputs needed for a particular land use type. Process models with a sound physical basis have a wider potential application than analogue models or expert knowledge. In particular, they are not restricted to the locality where they are developed. Also, by running the model for many years of climatic data and including the effects of other modelled processes like soil erosion, they can provide probability estimates of future yields or risks of low yields. This is a great advance over static methods of land evaluation that deal only with the situation at the time of the evaluation.

At present, many of these simulation models are complex, yet their physiological base is weak. Their development and maintenance consume many man-years of research time in well-found institutions, and their thirst for quantitative data on specific land qualities cannot yet be quenched by the data usually available from natural resources surveys. As a consequence, they are overly dependent on available climatic data.

There is a need for innovative research into simpler yet realistic models. There is also a need for surveys to provide quantitative data on relevant individual soil characteristics (like texture and soil depth) rather than taxonomic units; topographic characteristics (like slope angle, roughness and length of slope) rather than land systems or landform units; specific agro-meteorological and hydrological characteristics, and so on.

The present state of the art in simulation modelling is comparable to that in geographic information systems. They are both active and exciting research fields needing scarce and costly staff and equipment. Linking them should greatly increase their applications in land use planning: data can be re-analysed and presented anew almost instantly when conditions change, or to meet specific needs, and to incorporate estimates of risk into assessments (eg for cropping in areas with marked climatic variability).

## Financial and economic evaluation

If decisions are to be taken on rational grounds, decision-makers must weigh the natural resources information against economic and social imperatives. Usually, the balance is struck intuitively according to the information available to, and understood by, the decision-maker at the time. The technical difficulties of understanding, handling and combining large amounts of diverse data have severely limited the use made of natural resources information.

---

**Box 2.4** *Financial and economic evaluation of projects*

These evaluations are based on the assumption that prices reflect values, or can be adjusted to do so. Financial analysis (from the point of view of the individual land user or investor) and economic analysis (from the point of view of the wider society) are most appropriate where there is general agreement on values and policy goals, and where there are no unintended or off-site impacts such as soil erosion, salinity or loss of biodiversity.

From the point of view of the land user, the profitability of any activity on any particular patch of land is measured by *gross margin analysis* as the income from the activity less the production costs and overheads. This is a straightforward way of evaluating land under various kinds of management.

Where capital costs are substantial or where an economic evaluation from the point of view of society is required, more sophisticated *cost–benefit analysis* may be applied. The initial cost of development is set against the stream of future benefits (and costs) which are reduced to their present value by discounting (approximately the reverse of interest) to give three measures of worth:

1   *Net present value:* the present worth of benefits minus the present value of costs;
2   *Benefit–cost ratio:* the present value of benefits divided by the present value of costs;
3   *Internal rate of return:* the rate of discounting at which the present value of benefits becomes equal to the present value of costs.

Procedures for land evaluation are explained with examples by Dent and Young (1981). Discounting and calculation of these financial measures are explained in a manual by Price Gittinger (1982).

---

The power of economics in decision-making lies in its reduction of many variables to a single measure – money – and the general acceptance of a limited range of measures of project worth (Box 2.4). The rule is to undertake a project if the present value of benefits outweighs the present value of costs. But natural resources are not easily condensed into simple financial terms, and this is one of the reasons why natural resources information has been neglected by decision-makers. Air, water and soil, for example, are often treated as free resources in economic planning. Environmental services like climatic and flood control are unpriced. This problem is now being addressed by the maturing discipline of environmental economics which seeks to develop economic measures of environmental resources and services and incorporate these measures into decision-making (Turner, 1985; Pearce, 1991; Turner et al, 1994).

A principle of the FAO framework for land evaluation is that comparison between land use and land should be in terms of *benefits* yielded with *inputs* needed. If this comparison is to be made in economic terms, then suitability classes have to be calibrated in terms of both measurable outputs, and the inputs needed to achieve these, and monetary values have to be assigned to these inputs and outputs. This has not been done often, although one of the earliest applications of the FAO framework by Young and Goldsmith (1977) in Malawi included an economic evaluation. Ideally, the economic context should determine the cut-off values used for each land quality that determines a suitability class.

There are unavoidable uncertainties in economic evaluations. Costs and prices are themselves ephemeral, and performance depends on management as well as land qualities. The effectiveness of management is difficult to forecast and its ability to cope with problems – physical, social and economic – is not assessed in land evaluation. These problems can be overcome to some extent if the data about the relatively stable land qualities are held separately from performance and cost/price data. Performance and economic appraisal may then be recalculated from updated information about costs, prices and inputs. For example, the ALES computer programme (Rossiter and van Wambeke, 1993) provides a framework for land suitability evaluation according to decision rules entered by the user, and computes the conventional measures of financial suitability. By this means, any new or revised parameter can be taken into account and the re-evaluation undertaken quickly.

## Strategic land evaluation

A somewhat different question is whether or not the land has the capacity to meet the demands for products and services, now and in the future. This relates to land use policies to cope with population growth, other changes in demand, climatic change and technological change. The question has been approached by comparing needs or production targets for commodities and services with the capacity of the land to satisfy them; measuring the degree to which needs may be met and the flexibility of land use options in meeting these needs. Several recent attempts at strategic evaluation of land resources have used multiple goal programming, eg to assess scenarios for development in Canada (Smit et al, 1984) and the European Community.

## LAND USE PLANNING

Land use planning has proved to be much more difficult than gathering the supporting data. Land evaluation provides a link between the basic data and their application in land use planning; but the step from land evaluation to land use planning is a big one. Planning involves weighing land use opportunities against the problems involved, generation of a range of land use options, and making choices between these options. Not only does planning demand a broader range of professional expertise than for land evaluation, but the decisions made are much closer to the lives of land users and, in the final analysis, are political as much as technical.

It is only at the stage of land use planning or, more specifically, the implementation of the plan that something actually gets done with land resources information. Land users have always made plans to meet their needs from the resources at their disposal. These plans have usually been informal, and limited in scale and scope, but they *have* been implemented. Until recently, formal land use planning everywhere has remained very much a technocracy. Plans have been drawn up in the offices of the responsible agencies, remote from the areas being planned and often without any involvement by the supposed beneficiaries. Where the responsibility to plan and the power to implement have not been in the same hands, it has always proved difficult to impose these plans. Maybe this is just as well.

The last decade has seen two substantial responses to the shortcomings of such plans, namely decentralization and participation, both trying to bring planning closer to the people who have to implement and live with its results. Decentralization remains hamstrung by the lack of local capacity to undertake the kind of planning previously attempted centrally. The challenge of integrating broadly based participation in the planning process is greater and not merely logistic. It involves accommodating often divergent interests and dealing with power issues. Consequently, it requires both new methods and a leap of imagination from both planners and governments, which we discuss in the following chapters.

Here, we briefly review some of the standard techniques of land use planning and give examples of their application.

## Sectoral plans

The sheer variety of intentions and scales of planning, and the complexity of the situation on the ground, mean that no one clearly defined method of planning is discernible. However, conventional approaches have developed within individual sectors, eg in water resources development, forestry, infrastructure development, etc. For instance, objectives for both wildlife conservation and development planning have been pursued through land use zoning – that is allocating or confining particular activities or uses to particular areas. In the case of wildlife conservation, most countries follow the categories of land defined by the World Conservation Union (IUCN, 1982), although the local criteria have to be drawn up to arrive at those categories and, hence, zones on the ground:

I     Scientific reserve/strict nature reserve;
II    National park;
III   Natural monument/natural landmark;
IV    Managed nature reserve/wildlife sanctuary;
V     Protected landscape or seascape;
VI    Resource reserve;
VII   Natural biotic area/anthropological reserve;
VIII  Multiple use management area/managed resource area;
IX    Biosphere reserve;
X     Natural world heritage site.

Once protected areas have been designated, management plans can be prepared for them: the plan document providing baseline information on, among other things, landscapes, fauna and flora, and including prescriptions for use covering zoning, visitor management, resource management (eg fire control, game cropping, hunting quotas), research, infrastructure, education, etc. The master plans for national parks and wildlife management in Malawi are an excellent example (Clarke, 1983).

In the forestry sector, many countries have designated areas in various categories (such as protected forest areas, forest reserves, plantations) and have prepared management plans for such areas. Legal categories have been defined for forests and forest lands by functions and by conditions (ITTO, 1993):

1    Protection forests:
     1.1 Protection forests on fragile land,
     1.2 Forests set aside for plant and animal species and ecosystem preservation,
     1.3 Totally protected areas;
2    Production forests;
3    Conservation forests.

Frequently, protected forests are also designated as protected areas under the IUCN protected area categories. There are no universally agreed criteria either for zoning or for preparing management plans. Usually, documents describe the forest types and percentage cover, the standing stock, management objectives, logging quotas, controls and plans for regeneration/replanting.

In the agricultural sector, several national or departmental manuals have been drawn up and these show strong family relationships. For example, the manuals of the Land Use Services Division – later the Land Use Branch (Zambia Department of Agriculture, 1977) in Zambia, and by Shaxson et al (1977) in Malawi, evolved out of earlier manuals prepared by the Department of Conservation and Extension (CONEX) of the Federation of Rhodesia and Nyasaland. In Zambia, for example, land use plans follow a standard format covering a description of the area, history, physical conditions, resources, population, communications, present land use, assessment of agricultural potential, proposals for land use, plot demarcation, bush clearing, extension programme, roads, water supplies, soil conservation, staff and housing and financial appraisal. Good examples are the North Nyamphande Settlement Scheme (Wilson and Bourne, 1971) and the Msandile Catchment Plan (Wilson and Priestley, 1974). Such planning is often effective within its own narrow terms of reference, so long as the planning agency is also the executive agency and has some freedom of action; and so long as there are few conflicts of interest.

It is striking, looking at many sectoral plans, to note that they are usually strong on description, even strong on prescription, but with no obvious link between these aspects. Presumably, the link is made intuitively. Only in a few cases is the procedure of land allocation explicit.

## Land allocation procedures

Land use planning has been most effective and, sometimes, successful in settlement of 'empty' land and in new plantation and irrigation developments.

In these cases, administrative or engineering concerns have been paramount. Limited natural resources information has been used to guide the physical layout of farms, roads and the water distribution system and in the choice of crops. A good example is the work by Hunting Technical Services (1979/80) for the Mahaweli irrigation development in Sri Lanka. Land suitability evaluation was designed for precisely this purpose – allocating land use to land, or land to land use, according to its physical compatibility: irrigation on irrigable land, forestry on steep land and so forth. Given: (i) the goal (maximum production, settlement of the greatest number of people on viable farms, maximum conservation or some compromise between these); (ii) the area of land to be allocated; and (iii) basic information on land suitability; allocation can be done intuitively. The planner must also have an appreciation of the constraints of labour, capital and other inputs.

The techniques of land allocation have been adopted from farm planning and small-scale engineering development for settler farms, often based on land capability classification. Usually, arable land is at a premium and the simplest land allocation procedure is to establish how much arable is needed and in what sized management units. All the best land is then allocated to arable in units of appropriate size and if still more land is needed or convenient management units have to be made up, land of the next best category is tacked on until the demand for arable land is satisfied. Then the next priority is addressed in the same way.

Intuitive allocation becomes increasingly difficult where more (and more-complex) decisions have to be taken, for example for many management units over a large, complex area and, also, when even small mistakes are costly. Perversely, it becomes more difficult the more information is to hand. The problem of dealing with a welter of detailed information has been addressed by mathematical programming techniques for optimal land allocation (eg Dykstra, 1984; Hazell, 1986). An example of a mature land allocation procedure is the LUPIS package, developed in Australia (Ive et al, 1985; Ive and Cocks, 1988; Kessell, 1990) which links mathematical programming and a geographic information system. The first step is to define all the objectives of the plan and what allocation of land would satisfy each of these objectives. If all objectives cannot be satisfied, trade-offs between them are made until an acceptable compromise is reached, according to the weighting given to each objective (see 'Multiple criteria analysis', page 49).

But here is the crunch. To whom must the results of land use planning be acceptable?

The planner may start with a clean sheet of paper but the land is not a clean sheet. It has been over-written many times. There are few 'empty' areas to be settled and management is not so straightforward when there are many independent decision-makers and management units. Furthermore, there are conflicts of interest within local communities, and between government and local people. Not least of the difficulties is the failure of professional planners and administrators to comprehend and respect these various and often conflicting goals. This is exemplified by the case of the Barabaig in Tanzania (Box 2.5). There are also difficulties in getting several agencies to work together

---

**Box 2.5** *The case of the Barabaig*

In 1970, in response to an expected increase in demand for wheat, and with financial and technical support from Canada, the government of Tanzania (through the National Agricultural and Food Corporation) initiated a wheat production project which occupies 100,000 acres of the Basotu plains, Hanang District (Figure 2.1). The scheme is highly mechanized and based on the mono-cropping of hybrid wheat varieties along the lines of prairie wheat farming in Canada.

The land now used for wheat is also used for dry season communal grazing by the Barabaig – semi-nomadic pastoralists who number more than 30,000 in Hanang District. Each household manages its herd to maximize production of milk, meat and occasionally blood. Maize is obtained through exchange or sale of livestock, and from shifting cultivation by households with the help of relatives and neighbours.

The Basotu plains are droughty and without permanent water supplies. Therefore, the use of these plains by the Barabaig involves trade-offs between the productivity and stability of grassland production in different areas, within the constraints of water availability and the incidence of tsetse fly. The sustainability of the pastoral system is critically dependent on a flexible response to changing patterns of resource availability; opportunistic land use that allows the exploitation of key areas (eg wet depressions, river and lake margins) at particular periods. The Barabaig have developed their own natural resource management strategy which includes seasonal grazing rotation, grazing management, tsetse-control measures (eg burning bushland), controlling resource access, and customary regulations to control degradation (eg banning settlements in certain areas).

Appropriation of land for the wheat project has denied access to *muhajega* grazing (depressions on the plains which provide important dry-season fodder). Despite the fact that some Barabaig were resident in the area appropriated and that the *muhajega* were a vital forage resource, this land was described as idle during the project assessment (Young, 1983). Also, traditional burial sites have been ploughed up, causing much resentment.

The loss of *muhajega* has forced the Barabaig to adopt a new grazing pattern and to rely more heavily on the remaining forage areas, particularly in times when they would otherwise be rested from grazing. As a result, they are more intensively used during the critical regeneration period. There is increasing pressure to make greater use of those areas which have low potential, for instance the rift escarpment which has shallow soils on steep slopes, and the tsetse-infested bushland. This will inevitably result in further soil erosion and reduced production. According to the Barabaig, cattle populations in areas adjacent to the wheat farms have declined by about 30 per cent over the past 7 years, while productivity of the remaining herds has declined because of the loss of land.

The mechanized wheat farming, with no provision made for soil conservation measures, has also caused considerable soil erosion and siltation of water courses.

Even on narrow economic terms, the wheat project assessment was inadequate on two important counts: it overestimated the potential economic returns of wheat production and it ignored the opportunity costs of reductions in land available to the pastoral system.

The enforced changes in pastoral land use have resulted in a range of environmental impacts: a decrease in perennial species on the plains, accelerated soil loss on the hills, and clearance of bushland. In the bottomland areas still accessible to the pastoralists, there has been puddling and erosion of river and lake margins through excessive grazing.

Source: Lane and Scoones (1991)

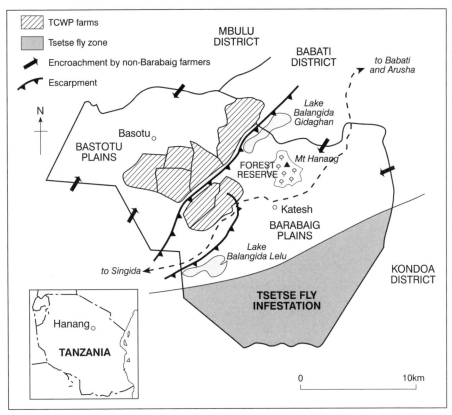

*Source:* Lane and Scoones (1991)

**Figure 2.1** *Map of Hanang District, Tanzania, showing major land use features*

and, often, a stark absence of technical solutions to land use problems that are practicable, profitable and easily incorporated into existing farming systems.

## Multiple criteria analysis

Where there are several goals and some of these are conflicting, the technique of multiple criteria analysis may be used to assess the degree to which these different goals may be attained. Van Mourik (1987) reports an early application in a livestock project in Tunisia but the greatest potential value of the technique may be in policy development. Policy-makers need specific information to develop policies and to identify what actions have to be taken to meet their various goals – for instance to reduce competition between rival demands, to meet present and future opportunities.

Interesting examples of an optimization procedure come from work undertaken by Dutch researchers using the multiple goal programming method of DeWit et al (1988) in dry parts of West Africa. They have been applied to decision-making at the household level, eg Maatman et al (1998), to optimize

---

**Box 2.6** *Optimizing land use in the Fifth Region of Mali*

The working steps in the Mali exercise were:

- Identify and characterize the land units (in this case, agro-ecological units defined by soil mapping units and climate).
- Identify and characterize promising land use types in terms of their land requirements, and their outputs given the constraints of each land unit.
- Scenario analysis. Several goals were defined that, in principle, could be optimized or assigned maximum or minimum values – production targets, monetary targets, risks, employment and emigration targets. Two technically feasible land use scenarios were built: one maximizing monetary value, the other maximizing self-sufficiency and both very different from the present pattern of land use.

The multiple goal programming technique was used to calculate the optimal pattern of land use and input required to achieve each selected goal.

*Sources:* Veenklas et al (1991); van Duivenbooden et al (1991)

---

and to assess the value of soil conservation practices; and at regional level by Veenklas et al (1991) and van Duivenbooden et al (1991) seeking ways to reduce competition between arable farming and livestock husbandry in the Fifth Region of Mali, to explore the technical possibilities of a more sustainable land use than at present, and to judge the economic viability of various land use options (Box 2.6).

There are two obvious reasons why little or no use has been made of such policy-support work in poor countries:

1   the huge data requirements and the substantial work needed to assemble the necessary data;
2   the complexity of the modelling and grade of facilities needed to operate it.

The optimization model for Mali has never actually been operated in Mali, so local planners have had no opportunity to use it!

There remain the tasks of translating model predictions into their social and economic repercussions, and then into the policies and action on the ground that are necessary to realize the predictions (or to prevent them). Modellers are supremely ill-equipped for this task, as is illustrated again in the example of Mali where the computed fertilizer needed to compensate for nutrient uptake by the modelled grain crop, for the Fifth Region alone, greatly exceeded the total imports for the whole country.

## Resource management domains

The planning techniques described so far come mostly from the land resources stable. Land resources specialists have analysed the potential for development and devised optimal solutions to problems of land resources in terms of biophysical variables (climate, soils, terrain, etc) and biophysical potential (eg crop yields). Their planning units have been biophysical land units. These are real areas of land but they are not areas of responsibility: they are not often decision-making units.

Administrators can work only with areas for which they have responsibility, which rarely coincide with those defined by nature. The same applies to land occupied and managed by individual land users or by whole communities.

The most effective decision-making unit depends on the problem in hand and, also, on who is making and implementing the decision. CIRAD (Centre for International Cooperation in Agricultural Research for Development), in West Africa, developed a distinctive geographic approach based on land use. Their basic decision-making unit, the *terroir*, is the village or social group territory in which farmers' practices are similar. Within the *terroir*, social, cultural and economic factors interact very powerfully with the biophysical environment to produce relatively homogeneous land use patterns – only providing that the pressures on the land are stable, which may have been the case in many rural societies until a couple of generations ago. The socially defined *terroir* is the basis of the participatory planning approach known as *gestion de terroir* (we discuss this approach in more detail in Chapter 3).

The concept of *resource management domains* is an attempt to combine both the socio-economic and biophysical appreciation of land resources. It arose from the concept of *recommendation domains* (Harrington and Trip, 1984) – groups of farmers whose circumstances are similar enough that they will be eligible for the same recommendation; and it was initially taken up by CGIAR[2] institutions looking for a more promising vehicle to enable the transfer of their high-flying technologies to wider areas and to people with different value systems.

The resource management domain is now transferred to a landscape unit as:

> '*A spatial unit that offers opportunities for identification and application of resource management options to address specific issues. It is derived from georeferenced biophysical and socio-economic information, and it is dynamic and multiscale in that it reflects human interventions in the landscape*'
> (Craswell et al, 1996).

In plain English, this means a patch of land in which the biophysical features are homogenous AND so is the management.

To date, resource management domains have been assembled from traditional biophysical land units (like soil mapping units) overlaid by socio-economic data. The various land units are linked with spatially referenced socio-economic attributes in a geographical information system (GIS), so that resource management domains can be identified by relating the spatial units to the issue being addressed. An alternative approach, based entirely on GIS, might be a raster system in which each grid square is characterized by both biophysical and socio-economic attributes, so that a resource management domain might be identified

---

2 CGIAR: Consultative Group of International Agricultural Research organizations.

by a digital elevation model and a decision-rule or algorithm for the problem in hand – though this is not an everyday procedure in poor countries.

Resource management domains lend themselves to computer-based information but the concept can be applied in a variety of situations without involving sophisticated technology. For land use planning, particularly for planning based on interactive participation of different stakeholders, the resource management domain can be a place where particular land resources are appreciated in a particular way, or are seen through particular eyes. Like a landscape painted by artists such as Constable and Turner, the eyes don't change the landscapes but their appreciation is different. So different ways may be found to achieve the same planning goals, depending on who is doing the planning.

## Land use planning experience in developing countries

Land use planning procedures range from the sophisticated to the summary. Examples of both have been tried in Tanzania where land use problems arose both from growing pressure on the land and from resettlement of people in new villages located by administrative fiat. The problems were addressed, first, by externally funded resource inventories, for example Mitchell (1984) in Tabora Region. Building on the Tabora surveys, a rigorous village land use planning procedure was developed based on algorithms of carrying capacity, economic viability, livestock carrying capacity and fuelwood availability (Corker, 1982).

The procedure proved too ambitious in terms of the time and expertise demanded (at least 45 days by an interdisciplinary team per village plan) and also in terms of the resources available to implement desired developments. With the benefit of this experience, a simplified two-stage procedure was evolved that demanded much less professional input. The simplified procedure comprised first, village-by-village appraisal of the land use situation by three-man teams with agronomic, livestock and forestry expertise; selection of priority villages; and then production of a framework plan in collaboration with the Village Council (Box 2.7).

Theoretically, this framework plan could then be fleshed out and implemented by the local community. But it wasn't. The local community was not involved from the outset (the initiative was external) nor in the active gathering and appraisal of data. Furthermore, Village Councils in Tanzania are essentially organs of government and Village Chairmen are political appointees. As a consequence, Council decisions do not necessarily reflect the needs, wishes or aspirations of the village community as a whole.

In Tabora, the planning teams were able to draw upon considerable expertise in land resources and planning built up by the externally funded land use project. In the absence of external funding, the cost of SPOT imagery to provide base maps and land resources information proved to be prohibitive, and elsewhere in Tanzania the natural resources expertise is not available.

Here we see two continuing problems of land use planning in the field:

---

**Box 2.7** *Procedure for framework village plans in Tanzania*

**I   Quick appraisal**

i   Using air photos or 1:50,000 enlargements of SPOT imagery, delineate village boundaries and measure areas suitable for cultivation and areas actually cultivated.

ii   Collect basic data on population and farming systems.

iii   Discuss local land use problems with village leaders to arrive at a crude appraisal of the match between village land resources and village needs.

This takes a three-man team about four days working in the village. On the basis of such appraisals, priority villages for the next stage of planning can be identified at a district level.

**II   Framework plan**

i   Sketch landforms, land use, soils, eroded areas, water sources and tracks on 1:50,000 imagery.

ii   Field check, especially of soils and water sources.

iii   Survey village and sample households to determine population distribution and growth, land holdings, livestock ownership, levels of production and other economic activities.

iv   Assess land suitability and draw up an indicative land use plan. Discuss its implications with the people concerned.

The whole procedure, including the production of a framework plan, requires about two weeks of work by a three-man planning team living in the village.

Implementation relies on devolution of authority to the Village Council which is supposed to resolve conflicts and determine the priorities for development. The resulting framework plan (Figure 2.2) is its responsibility. It can allocate land according to customary law, lay out individual farm plots, and manage communal land uses such as woodlots and grazing reserves. Locally developed plans can be implemented because they do not rely on major external inputs.

*Source:* Wheeler et al (1989)

---

1   A growing awareness of the need for people's participation in planning without grasping its far-reaching implications of this for political development and professional procedures.

2   An acute shortage of both professional expertise and of resources for both planning and implementation.

These problems may be tackled from opposite poles. One approach is to prescribe the detailed procedures to be followed, tailoring them to specific local circumstances. The alternative is to set out a route and the steps to be followed, leaving people on the ground to adapt the steps and fill in the details to suit their own circumstances.

An example of the first approach was developed in the 1980s in Bangladesh by Hugh Brammer in support of an attempt to revive the Thana[3] Development

---

3 Thana: a local self-government/administrative unit in Bangladesh, usually including 150–175 villages and with a population of about 200,000 people.

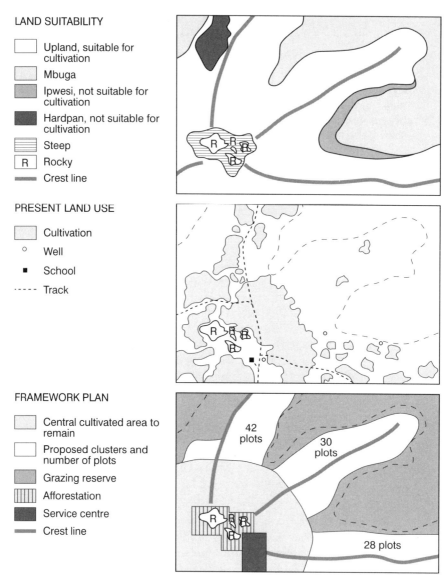

**LAND SUITABILITY**

☐ Upland, suitable for cultivation

☐ Mbuga

☐ Ipwesi, not suitable for cultivation

■ Hardpan, not suitable for cultivation

☰ Steep

R Rocky

▬ Crest line

**PRESENT LAND USE**

☐ Cultivation

○ Well

■ School

---- Track

**FRAMEWORK PLAN**

☐ Central cultivated area to remain

☐ Proposed clusters and number of plots

☐ Grazing reserve

▦ Afforestation

■ Service centre

▬ Crest line

42 plots

30 plots

28 plots

*Source:* Wheeler et al (1989)

**Figure 2.2** *Land suitability, present land use and framework plan for a village in Tanzania*

Programme. In the 1960s, this programme had made considerable progress in land drainage, irrigation and rural roads based on a participatory planning model developed by Akhter Hamid Khan at the Comilla (now Bangladesh) Rural Development Academy. Recognizing that officials – mainly urban – often knew less about local environments than the farmers they were supposed to be talking with, Brammer wrote a training manual. Its aim was to bring officials on to an equal footing, to make dialogue possible, by enabling them to make use of soils and land use information that had been gathered by surveys between 1963 and 1975.

The manual gives detailed instructions for every step and calculation: for example, to assess the area of land available, local food sufficiency, allocation of water for irrigation, even dealing with contingencies, and the procedure was dovetailed with the many layers of local administration and to the responsibilities of officials at each level.

Although this approach avoids the problem of a lack of local technical expertise by working out all the answers beforehand, its strengths are also its weaknesses. It would be difficult to change the procedure from below to deal with new questions and, while the intention is to engage local people in dialogue, it involves consultation rather than active participation by local communities (see Table 3.1). More empowering approaches to participatory planning are discussed in Chapter 3.

This particular planning initiative failed for a number of reasons. Possibly the main one was the fall of the government that promoted it, just as the original Thana development planning programme foundered during and after the separation of Bangladesh from Pakistan. Without strong political support, it fell prey to inter-agency rivalries. A subsequent government stripped Thana Councils of development powers and funds, and established government departments continue to operate their independent area development/ technology transfer programmes, and refuse to be coordinated.

All-too-human factors also came into play: the threat perceived by senior officials in the Extension and other departments from their junior officials learning techniques they did not themselves understand, and the fact that dialogue remains an alien concept in the culture of officials throughout the subcontinent. A revised *Manual of Upazilla Planning*, prepared for use by extension officers, never got past the discussion stage and no further attempt has been made to revitalize the integrated planning initiative.

## FAO guidelines for land use planning

An alternative model of planning evolved in the course of a series of expert consultations convened by FAO. The *Guidelines for Land Use Planning* (FAO, 1993, Figure 2.3) outline a ten-step procedure that may be applied at any scale of planning, from global to the individual farm, but leaves the fine detail to the people on the ground.

The thrust of the spear point in Figure 2.3 implies steady progress from one step to another but, in practice, we often have to retrace our steps to take account of new information, changing conditions or new goals.

The crucial question is 'Who shall be responsible for each step?'. Here the FAO guidelines betray their expert lineage by allocating most responsibilities to the professional planning team. Consultation is implicit in Step 1 (establish goals and ground rules) and, through rapid rural appraisal, in Step 3 (structure the problems and opportunities). Public and executive discussion is explicit in Step 7 (choose the best option – from among those already identified and appraised by the professionals!). Then, at Step 9, the plan is supposed to be made flesh by the beneficiaries.

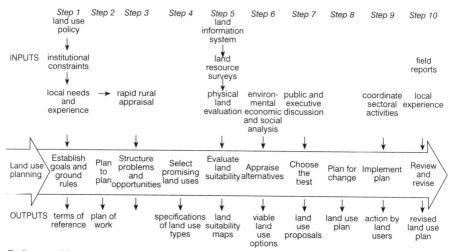

**Figure 2.3** *Steps in land use planning*

### Define problems
*Step 1:* Establish the present land use situation and the needs of all the stakeholders; agree and specify what you want to achieve
*Step 2:* Plan to plan. Organize the work needed
*Step 3:* Structure the problems and opportunities. Socio-economic and biophysical considerations should be given equal weight

### Model solutions
*Step 4:* Devise a range of responses. Identify or design alternative land use types or patterns of land use that might achieve the goals
*Step 5:* Evaluate land suitability. For each promising land use type, establish its land requirements and match these with the qualities of the land
*Step 6:* Appraise alternatives. For each well-matched combination of land use and land, assess its environmental, economic and social impact

### Decision
*Step 7:* Choose the best achievable land use
*Step 8:* Draw up a land use plan, allocating land to land use and making provision for appropriate management

### Test solutions
*Step 9:* Implement the plan. Action by implementing agencies and land users
*Step 10:* Learn from the plan. Monitor progress. Revise the plan in the light of experience and to accommodate new goals

*Source:* Dent (1988)

A checklist of the activities required at each step is provided and, in principle, these steps can be taken either by top-down or bottom-up planning. In practice, however, few established institutions or individual professionals have been able to break away from their paternalistic tradition.

# Faith in negotiation

Plans handed down from above have a consistent record of failure:

- *Failure to address all the issues.* There are many instances of concentration on production at all costs. On the other hand, efforts at conservation typically deal in isolation with technical measures such as destocking of grazing lands, or construction of earthworks to retain soil and water. The social and economic imperatives that drive unsustainable use are not dealt with.
- *Failure of information.* Lack of data, failure to make use of either detailed local knowledge or relevant technical knowledge.
- *Failure to integrate efforts and goals.* Typically we see competing projects and competing institutions with disparate goals.
- *Failure of institutions.* Failure to cope when situations are complex or when times are hard, and a reluctance by governments and development agencies to address the problems of institutional weakness.
- *Failure to address the legitimate goals of all the stakeholders*, to involve all the stakeholders in planning the use of land, or to empower them to manage resources in common. Fiats imposed from above are resented, resisted, ignored and, ultimately, overturned.

Therefore, there has been a general failure of implementation of plans, at least in anything like the shape envisaged by the planners.

Attempts to learn from these failures have not yet advanced beyond statements of principle. FAO has tried to develop a new approach to land management and land use planning which, it claims, emphasizes the integration of physical, socio-economic and institutional aspects of land use, as well as the need for the active participation of stakeholders in decision-making (FAO/UNEP, 1997). This latest expert response involves few changes to the existing land use planning guidelines and has yet to be field tested (a pilot exercise initiated in Sierra Leone had to be aborted due to civil war) but it does emphasize two principles that may serve as a basis for more effective land use plans:

1   First, successful implementation of a plan depends upon common ownership of the problems and the proposed solutions by the people who will be affected. This common ownership may come from consensus about the goals and the actions needed, or from a negotiated compromise between groups with different goals and different insights.
2   If there is to be negotiation, there must be some forum that commands general respect where the parties can negotiate – to agree goals, prioritize problems, contribute their knowledge, and allocate responsibilities for action (see 'Better institutions to make rural planning and development work: possible ways forward' on page 167). Of course, the necessary readiness to negotiate a compromise may not be present, but no better alternative has been proposed.

Holding to these principles, Dalal-Clayton and Dent (2001) have redefined some of the steps in land use planning (Figure 2.3) with a further step at the very outset, a Step 0, to set off on the right foot (Box 2.8).

**Box 2.8** *Redefined steps in land use planning*

### Step O: Foster partnership

Legitimacy and, therefore, broadly based commitment to a plan requires, first, identification and involvement of the stakeholders. A stakeholder is anyone, or any group, with an interest in or affected by the issue or activity. Where natural resources are concerned, everyone is a stakeholder but their commitment, or opposition, depends on the benefit they stand to gain, or forfeit.

Different stakeholders have different interests and, often, these conflict. Unresolved conflicts are the root of much land degradation because they lead to short-term exploitation of resources; so a prerequisite of sustainable management is a forum where all the stakeholders, or their representatives, meet on equal terms to resolve conflicts of interest or negotiate an acceptable compromise. Very often, there are existing institutions – maybe a local water management group, cooperative, or assembly – that can be empowered to take on a wider responsibility. If there is no forum, one must be created and experience shows that such institutions work best when the rights of resource users to manage the resources on which they depend are respected by outsiders and governments (Ostrom, 1990).

### Step 1: Setting the goals and ground rules

This is the first item for negotiation. The goals may emerge from local concerns or national issues. They must be specified and, since it is likely that there will be several, they must be ranked. The criteria by which decisions will be made must also be agreed and, since they too will be several, they must be weighted.

Decisions have to be made about the location, size and boundaries of the planning area; the scope and time frame of the plan; the responsibilities of the various partners, eg in providing funding, labour and facilities. All of these require basic information about the land and its resources; the people and their needs, rights and responsibilities; and the organization of administration and services. And if negotiations are to be meaningful, all the stakeholders must hold this information and be equipped to make use of it.

### Step 2: Organize the work

List the tasks to be accomplished; the people and other resources required; and draw up a schedule of activities. Depending on the size and complexity of the planning operation, the assembly of data and preparation of proposals may be delegated to professional staff but the decisions should be made jointly by all stakeholders – not by government agencies acting alone.

### Step 3: Structure the problems and opportunities

A problem is a gap between the present situation and some preferred state of affairs. The essential activities of this step are to find out about the present situation; judge ways in which it is bad; and identify ways in which it might be made better (Siffin, 1980). All participants will have their own ideas about the nature of the problems and the opportunities for change, and these must be canvassed. Specialist research may be required to establish the facts. Table 2.3 lists examples of the kinds of problems that land use planning is expected to solve and the broad-brush solutions that might be proposed at Step 3.

### Step 4: Specify alternatives

Usually, there are several ways of tackling a problem that are worthy of rigorous appraisal before the best option can be chosen. These may include non-land use options; projects to accomplish specific tasks such as construction of bunds and well boring; education programmes; and the adoption of new or improved land use systems. In terms of conventional land evaluation, this step includes the identification and definition of promising land use types.

### Step 5: Evaluate land suitability

This step involves a comparison of the requirements of the promising land use types with what the land has to offer. In a large project, this biophysical evaluation is usually undertaken by land resources specialists.

### Step 6: Appraise the alternatives

The alternatives are appraised in terms of their economic, environmental and social consequences. In the past, appraisal has been confined to one question – will it pay? Among the ground rules for planning that have to be negotiated at Step 1, and revisited at this step, are the criteria by which the alternatives will be judged; so that supporting information may be assembled.

### Step 7: Choose the best option

Now the process moves firmly back into the hands of the decision-makers. Although there may be shuttling between them and any decision-support team to answer sequences of 'What if?' questions or to test modified proposals, the decisions are likely to be the outcome of intuitive judgement and hard bargaining.

### Step 8: Prepare a land use plan

This will state what has been decided and why, how it should be accomplished and, if funding is required, how much it will cost. Where a formal written plan is required, this will draw upon the preparation work of earlier steps, especially the natural resources studies of Steps 3 to 6 but, beyond this, substantial logistic and financial planning will be involved.

### Step 9: Implement the plan

Most failures of planning have failed in the doing, most often through lack of support which should already have been evident through Steps 0, 1, 3 and 7. For a large project, implementation may require an institution of its own with time and a budget orders of magnitude greater than the preparation stages, whereas an independent farmer or a local community may implement their own plan in their stride, with their own resources.

Where a project involves government departments, new project management institutions should *not* be created merely in order to circumvent the legitimate official structures, even if they are ineffective. This approach has often been followed in the past, particularly for large donor-funded projects. The project might operate effectively while donors are involved, but experience shows that the new institution and the project usually collapses once donor support ends. The better option is to work within the existing government or other structures and try to overcome any institutional problems.

---

**Step 10: Monitor and revise**

Major projects establish regulatory procedures to ensure that money is being spent as intended and work is completed to the required standard. Of course, this is important but it is also necessary to know if the goals of the plan itself are being attained, eg arrest of land degradation, improvement of water supplies and water quality, maintenance of biodiversity. The original plan will not be perfect and changing circumstances may also demand a change of direction, so information on relevant natural resources indicators must also be gathered and fed back into the planning system.

---

Many of these steps are less complicated where the area and scope of the plan is restricted, and the decision-makers themselves have close links with each other and with the land. But whatever body takes on the job of land use planning, it still needs the input of natural resources information, homespun or professional. Mountains cannot be moved, even by negotiation, unless the existence and whereabouts of the mountains are known about in the first place.

**Table 2.3** *Examples of land use problems and responses*

| Symptoms | Underlying causes | Response options |
|---|---|---|
| Migration to towns | **Social:** | Reform or clarification of land |
| Low rural incomes | • population pressure | tenure |
| Lack of employment | • inequitable access to | Provision of credit guarantees |
| opportunities | land, capital | Marketing and infrastructure |
| Shortage of land, fuel, timber | • lack of infrastructure | initiatives |
| Encroachment on forest and | | Improved irrigation |
| wildlife resources | **Natural hazards:** | Fertilizer supply |
| Land degradation | • inadequate water | Integrated pest management |
| | • droughty soils | Agroforestry |
| | • rough terrain | Soil and water conservation |
| | • flooding | |
| | **Mismatch between land and land use:** | |
| | • clearance of forest on steep lands | |
| | • inadequate bush fallow | |
| | • inadequate soil and water management | |

## IMPACT ASSESSMENT

Over the past 30 years, a battery of techniques has emerged for the environmental assessment of projects, programmes and policies. They include environmental impact assessment (EIA), risk assessment, social impact assessment, and the routine desk-screening of proposals for development. A useful, up-to-date summary of these techniques is provided in Donnelly et al (1998).

Since its introduction in 1969 in the USA under the National Environmental Protection Act, more and more countries have introduced legislation that requires EIAs for certain categories of development. EIA predicts the likely environmental impacts of development projects, identifies ways to reduce or mitigate unacceptable impacts, and presents these predictions and options to decision-makers. However, EIA has been essentially reactive – often conducted after the design phase of projects and, sometimes, even after implementation (for instance, the EIA for the Victoria dam in Sri Lanka). Current practice deals mainly with environmental impacts in biophysical terms. There is, however, a widening recognition of the need to integrate, within the EIA process, the consideration of social, economic and biophysical aspects. Increasingly, EIA is being seen as an important tool for involving different groups of stakeholders in the development process, and as a means of ensuring the accountability of development proposals.

There is also growing recognition of the need for environmental management and for introducing environmental considerations at earlier stages of the decision-making process. EIA has rarely been attempted at policy or plan level, although this approach is now being promoted through strategic environmental assessment. Dalal-Clayton and Sadler (1998a; 1998b) provide a review of its application and potential in developing countries.

There tends to be a stand-off between EIA and other statutory and non-statutory rural planning instruments, such that the one rarely informs the other and EIA studies can be ignored. There is also a tendency in many countries (eg Zimbabwe) for developers to hire consultants to undertake EIA studies, which are then prejudiced in favour of the priorities of the developer, but given a patina of professionalism. The State, which should act as an arbiter in these matters, often has limited resources and so the EIA studies are not seriously examined or questioned. A recent review of the performance of EIA in Tanzania (Mwalyosi and Hughes, 1998) demonstrates how slight the impact of EIA is on decision-making and how rarely it has led to design modifications.

## DECENTRALIZED DISTRICT PLANNING

District planning has long been a feature of regional planning. As for land use planning, many countries have prepared manuals for district-level planning, and these usually concentrate on the issues to be covered rather than how to do it. The top-down nature and project/programme orientation of district plans throughout the 1970s and 1980s and even in the 1990s is typified by that of the Akuapem North District Development Plan, Ghana (Box 2.9). In many countries, eg Nepal, district plans are sectorally based, lack integration and still represent little more than a catalogue of projects, initiatives and events that might take place in the future, providing funding is available from central government or other sources.

As already noted above, district authorities in most developing countries have long suffered from a lack of institutional capacity to prepare, implement and monitor plans. While donors have supported some projects in district plans,

---

**Box 2.9** *Akuapem North District Development Plan, Ghana (1996–2000)*

This plan was prepared by the Akuapem North District authorities within the framework of planning guidelines issued by the National Development Planning Commission which prescribes format and content. District plans in Ghana are expected to be based on the goals and objectives of the 25-year Perspective Plan: Ghana Vision 2020, The First Step: 1996–2000.

**Contents**

*1 Executive summary*
• General stagnation of population growth, inadequate social infrastructure, low incomes, weak settlement hierarchy, and weak administrative and management capacity.
• The focus of the development plan – stated as the promotion of social and economic growth and development of the district consistent with the objectives of Ghana Vision 2020.
• Strategies for achieving the development objective.
• Specific development programmes following from the strategies.
• Implementation of the 5-year district development plan.

*2 The existing context*
• Covers: social services, economic base, infrastructure, environment, land and land management, housing, urban management, and NGOs.

*3 Medium-term development proposals*
• Covers: development constraints and opportunities, development objectives and focus of the district development plan, development programmes and projects.

*4 Implementation of the 5-year development plan*

*5 Monitoring and evaluation*

*Source:* Botchie (2000)

---

implementation of plans has been hampered by less-than-enthusiastic support from central government agencies and by shortages of finance and resources available to district authorities. The latter bottleneck is increasingly being recognized. For example, in Zimbabwe, the decentralization model is seeking to enhance the role of Rural District Councils (RDCs) in planning and managing districts. The vision is that the RDC should act as the pivotal point for all rural planning initiatives. A Rural District Development Committee, led by the RDC but embracing all relevant district stakeholders, is required to produce strategic and three-year rolling plans to guide development and set priorities. The Rural District Councils Capacity-Building Programme aims to build the capacity of the RDC to lead this process and become a strategic planning authority (see Figure 2.4), although progress with the programme has been interrupted due to withdrawal of donor support in 2001 as a result of the political situation in the country.

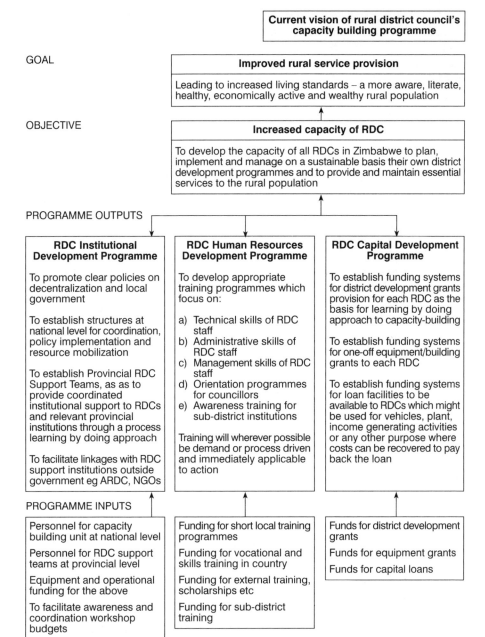

**Current vision of rural district council's capacity building programme**

GOAL

**Improved rural service provision**

Leading to increased living standards – a more aware, literate, healthy, economically active and wealthy rural population

OBJECTIVE

**Increased capacity of RDC**

To develop the capacity of all RDCs in Zimbabwe to plan, implement and manage on a sustainable basis their own district development programmes and to provide and maintain essential services to the rural population

PROGRAMME OUTPUTS

| **RDC Institutional Development Programme** | **RDC Human Resources Development Programme** | **RDC Capital Development Programme** |
|---|---|---|
| To promote clear policies on decentralization and local government | To develop appropriate training programmes which focus on: | To establish funding systems for district development grants provision for each RDC as the basis for learning by doing approach to capacity-building |
| To establish structures at national level for coordination, policy implementation and resource mobilization | a) Technical skills of RDC staff | To establish funding systems for one-off equipment/building grants to each RDC |
| To establish Provincial RDC Support Teams, as as to provide coordinated institutional support to RDCs and relevant provincial institutions through a process learning by doing approach | b) Administrative skills of RDC staff<br>c) Management skills of RDC staff<br>d) Orientation programmes for councillors<br>e) Awareness training for sub-district institutions | To establish funding systems for loan facilities to be available to RDCs which might be used for vehicles, plant, income generating activities or any other purpose where costs can be recovered to pay back the loan |
| To facilitate linkages with RDC support institutions outside government eg ARDC, NGOs | Training will wherever possible be demand or process driven and immediately applicable to action | |

PROGRAMME INPUTS

| | | |
|---|---|---|
| Personnel for capacity building unit at national level | Funding for short local training programmes | Funds for district development grants |
| Personnel for RDC support teams at provincial level | Funding for vocational and skills training in country | Funds for equipment grants |
| Equipment and operational funding for the above | Funding for external training, scholarships etc | Funds for capital loans |
| To facilitate awareness and coordination workshop budgets | Funding for sub-district training | |
| Experimental training budgets | | |

*Source:* PlanAfric (2000)

**Figure 2.4** *Current vision of the Rural District Councils Capacity-Building Programme, Zimbabwe*

Some decentralization programmes have provided direct government funding to district authorities to implement their plans. In Guatemala in the late 1980s and early 1990s, 8 per cent of the national budget was distributed directly to municipalities, based on their population. But this structure was abandoned by the next government after two years. The new decentralized planning system in Ghana also passes on a guaranteed proportion of government revenue to the districts (see Box 2.10). But there are still institutional and capacity problems. In 1995, each district in Ghana was given only three months in which to adopt new planning guidelines from the National Development Planning Commission and prepare district development plans. Given the shortages of available information and of experienced professional staff, all 110 districts appointed consultants to prepare the district development plans in 1995. A recent study of five of these plans (NRI/UST, 1997) found that they contained district profiles which were very descriptive rather than analytical:

> *'There is no attempt to address the identified problems and constraints by putting forward alternative strategies or choices, or to bring together various sectoral proposals into an integrated co-ordinated plan. None of the five plans have (sic) a clear vision about what trends are likely to occur over the 5-year time period or what broad longer-term spatial trends would be appropriate for their area. Consequently, each district ends up producing a series of project proposals for implementation during the 5-year period, but these are often conceived independently of the goals and objectives or analyses of current problems.'*

This diagnosis typifies the state of district planning in developing countries. Furthermore, the plans tend to concentrate on the provision of social and economic infrastructure, particularly on education, health or public latrine projects. Analysis of land capability has not been undertaken and maps are not available. Settlement planning is absent and none of the plans put forward an environmental management strategy. Community participation, seen as a priority in the planning guidelines, is weak. At best, in one of the five districts, workshops were held for assembly members, heads of departments, chiefs and other opinion leaders.

The NRI/UST (1997) study revealed a continuing gap in the linkages between district level planning and village planning in Ghana – a major problem in most developing countries:

> *'Town or village development committees do not appear to have been actively involved in village planning and doubt exists about the degree of consultation which is taking place in most villages. The traditional authorities appear to be making ad hoc short-term decisions about the sale of land in advance of the preparation of any village plan and often without considering the long-term consequences or implications on the environment. Currently, the district town and country planners are handicapped in offering development control advice or preparing layouts at the village level because they lack information both about the activities of the chiefs in releasing land for development, as well as on housing and other constructions which have been developed.'*

**Box 2.10** *Ghana's new decentralized planning system*

Previously, government planning in Ghana was highly centralized and marginalized local government. Recently, a decentralized planning system has been introduced which is participatory and bottom-up (see Figure 2.5). Each of Ghana's 110 districts now has full responsibility to develop and implement its own medium-term (5-year) and annual development plans. This responsibility lies with the District Assemblies (DAs) which are either metropolitan (population >250,000), municipal (one town with a population >95,000) or district (population >75,000). Members of DAs are elected every four years from people ordinarily resident in the district. In addition, the President, in consultation with traditional authorities, can appoint one-third of the members – in practice from individuals nominated by the districts to ensure the inclusion of people with skills and expertise so that the business of local government is properly conducted. The District Chief Executive (DCE) is also appointed by the President and is subject to approval by the DA. A District Planning Coordinating Unit acts as the secretariat to the DA for planning purposes.

The National Development Planning Commission (NDPC) has prepared guidelines for district-level planning (NDPC, 1995). Each district plan must be subjected to a public hearing. District development plans are harmonized at the regional level by Regional Coordinating Councils, and these regional plans are then consolidated with individual sector plans (prepared by line ministries and also subject to hearings) by the National Development Planning Commission into a National Development Plan. The NDPC undertakes this task through cross-sectoral planning groups which have representatives from the public sector, business, university, districts, scientists, trade unions and farmers. The country has implemented the first of these rolling medium-term development plans covering the period 1997–2000 (Ghana NDPC, 1997).

Through new financial arrangements under the Constitution, at least 5 per cent of internal government revenue is allocated by parliament to the District Assemblies Common Fund. Each year, parliament agrees a formula for the distribution of the fund. This takes into account the population of each district, weighted by its development status (judged pragmatically by indicators of socio-economic development such as the number of pupils attending school, the presence of a commercial bank) and its revenue mobilization effort (the percentage increase over the amount collected in the previous year). DAs are able to use these funds for capital expenditure of a development nature within the district plans. For example, at present, five districts are implementing a poverty reduction programme and the targeted communities decide upon what action is to be taken.

The base level for planning lies in Unit Committees for settlements or groups of settlements with a population of 500-1000 in rural areas and 1500 for urban areas. Community problems are identified here and goals and objectives set out and passed up through higher level councils to the DA. Committees of the DA consider and prioritize problems and opportunities. Departments of the District/Municipal/Metropolitan Assembly together with sectoral specialists, NGOs and other agencies collaborate to distil the ingredients of the district plan. The District Planning Coordinating Unit integrates and coordinates the district sectoral plans into long-term, medium-term and short-term with annual plans/budgets for consideration by the DA.

While legislation has given power to the DAs, there is limited capacity to undertake these new responsibilities and the DAs have also been pressured by line ministries to establish a variety of district committees, eg for environmental management, disasters, health, etc. According to the Ministry of Local Government and Rural Development (Ghana MLGRD, 1996), districts still lack decentralized departments of many line ministries. In the past, line ministries operated centrally determined programmes. Under the Local Government Services Bill, district offices of line ministries will become departments of the DAs answerable to the DCE, but there is resistance to the new arrangements from the ministries; there is lingering allegiance of district level staff to their regional and national headquarters; and some staff have been reluctant to accept postings, particularly to deprived districts. The quality of district staff is relatively poor although it has improved following training and refresher courses, and postings of national/regional staff to districts. As a result, some DAs have turned to consultancy firms to assist in the preparation of their plans. The new system also faces continuing logistical problems (inadequate office and residential accommodation, equipment, vehicles, etc).

District planning officers have received some training to facilitate community meetings through which communities can identify their concerns and needs. This process has led to some unexpected requests. One village, for example, wanted funds for a brass band, reasoning such a band at their weekly village market would attract people from other villages and enhance economic growth.

### Aims in the short, medium, and long-term

Placing the focus of the new planning system on the districts provides an unprecedented opportunity for local communities to participate effectively in the conception, planning and implementation of local development.

In the *short term*, the aim is to put the new planning process into effect. Its goals are:

- restructuring of the political and public administrative machinery for development decision-making at both national and local levels;
- organization of spatial development to attain functional efficiency and environmental harmony;
- integration of local government and central government at the regional and district levels;
- decentralization of the development planning process;
- adequate transfer of financial, human and other resources from central government to local authorities.

Significant progress has already been made, eg:

- Clear objectives for decentralized planning and a structure for this have been established.
- The necessary legal instruments have been put in place, notably the Local Government Act 462, 1993. This instrument conferred legislative, executive and deliberative powers on the DAs. It did not represent a shift against those with more power, but gave legal authority to the DAs to promote the development of their districts to ensure equity and the reduction of poverty and vulnerability.
- The establishment of the District Assemblies Common Fund and ceded revenues.
- Training to improve human resource capacity for DAs.

In the *medium term*, further progress towards achieving these aims is expected through the implementation of the National Medium-Term Development Plan and the 5-year district development plans (both in progress).

The *long-term* perspective is provided by the National Development Policy Framework (*Ghana Vision* 2020) and both the National Medium-Term Development Plan and the District Development Plans are meant to be consistent with this framework.

*Sources:* George Botchie (2000 and pers. comm.); Dalal-Clayton and Dent (2001)

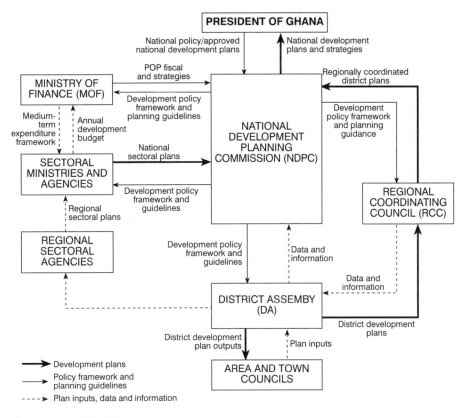

*Source:* Botchie (2000)

**Figure 2.5** *The new planning system in Ghana*

Echoing points made in previous studies, eg Dalal-Clayton and Dent (1993), the study concludes that, for forward planning, up-to-date information is required (eg about which villages have been growing or declining, and on the distribution or use of natural resources) and that better natural resource management requires that the necessary information is both accessible and usable for planning and management purposes.

In sub-Saharan Africa, and particularly in the wildlife sector, Namibia has experimented with innovative approaches to decentralization through processes of empowerment and participation, but again there are constraints at all levels of the planning process related to the mismatch between national frameworks and local realities.

## SOME PLANNING RESPONSES TO THE CHALLENGE OF SUSTAINABLE DEVELOPMENT

The Report of the World Commission on Environment and Development (WCED, 1987) (the Brundtland Commission) emphasized that sustainable development could not be achieved without the active participation of local communities in the development process, including the management of natural resources.

The issues to be addressed in responding to this challenge clearly include good governance, equitable access to resources and institutional coordination.

### Techniques

Our perception of the present situation and of the tasks ahead is built up from a range of techniques. Figure 2.6 plots some of these techniques in relation to the degree of participation – for example GIS and remote sensing score low on participation while participatory rural appraisal absolutely depends on it.

Taken together, they form a framework that has been called sustainability analysis (Dalal-Clayton, 1993).

### National and regional planning exercises

The means to link planning at national and local level and to actively involve local communities have been slow to develop. The publication of the World Conservation Strategy (IUCN/WWF/UNEP, 1980) stimulated many countries to prepare cross-sectoral National Conservation Strategies to integrate environmental concerns into the development process. National Environmental Action Plans, initiated by the World Bank in 1987, also involve reviews of the natural resource base and environmental problems. On a sectoral basis, many countries have prepared National Forestry Action Plans and water master plans. International response to catastrophic flooding in Bangladesh in 1987 and 1988 led to the Bangladesh Flood Action Plan, funded by a wide range of donors and coordinated by the World Bank. It was launched in 1990 as a series of regional and supporting studies aimed at identifying appropriate action. Despite its promotion as a comprehensive exercise, the regional planning studies which have emerged have focused mainly on flood control. They have not dealt with inter-regional issues (and vast amounts of water move through Bangladesh!); have not captured the wealth of historical experience in Bangladesh concerning water management; have involved very little effective participation of the people the plan aims to protect; and have suffered from a lack of reliable baseline data (Hughes et al, 1994).

### Sustainable development strategies

*Agenda 21* (UNCED, 1992) urges all nations to develop a national sustainable development strategy (NSDS) to implement, at the national level, its priorities and recommendations. Countries have responded to this challenge in a variety of ways.

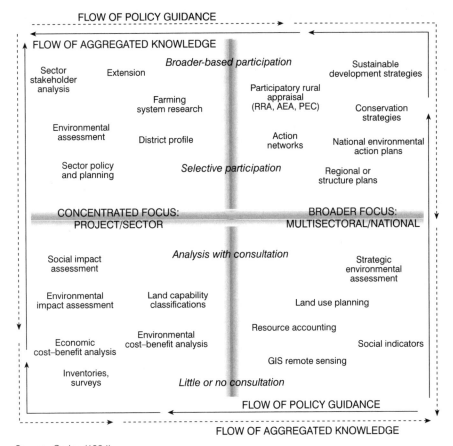

*Source:* Carley (1994)

**Figure 2.6** *Techniques of sustainability analysis and participation*

In most countries, there is a range of past and existing strategic national and decentralized planning approaches. Many of those in developing countries have been externally conceived, motivated and promoted. Few of them have adopted or built on the systems, processes and practices that have operated in the country for some time. Very few countries have developed a specific or overarching strategy for sustainable development, and specifically labelled it as such, and indeed, this is not necessary. It *will* be important to build on what exists: starting by identifying the existing processes and initiatives in the country. Below, we provide a brief overview of existing country strategic planning frameworks.

## National strategies

There is a strong tradition in most developing countries of preparing periodic national development plans, typically covering a five-year span. Usually, line ministries prepare sector chapters following guidance issued by a national coordinating body. Such plans tend to set out broad goals and include projects and activities to be funded from the annual recurrent and development budgets. Economic, or occasionally social, imperatives have been predominant. These plans tend to be linked into the annual budget or to the medium-term expenditure framework – a three-year rolling budget process.

In the past, there has been little involvement of civil society or the private sector in developing or monitoring such plans. But there is increasing evidence of stakeholder participation in these processes in a number of countries. There is also greater use of environmental screening mechanisms (although usually to screen out certain bad impacts, rather than to optimize environmental potential).

Connected to these planning instruments, line ministries prepare sector-wide plans and investment strategies – transport, agricultural, health and education strategies and so forth. Many countries also prepare cross-sectoral strategies. Examples include strategies for reducing HIV/Aids or for improving rights for women. Cross-sectoral environmental strategies include coastal zone management plans, and the Bangladesh Flood Action Plan in the early 1990s, which led to the more participatory development of a National Water Plan. Many environmental strategies respond to the Rio Conventions, such as Biodiversity Action Plans and National Action Plans on Combating Desertification. Responsibility for the preparation of these plans has often been given to environment ministries. Similarly, national forest programmes are in progress to implement the proposals for action of the Intergovernmental Panel on Forests.

Some governments responded to *Agenda 21* by giving renewed attention to, or building on, the mainly environment-focused national conservation strategies and national environmental action plans that were developed in the 1980s and early 1990s. Subsequently, a range of countries have prepared national Agenda 21s to set out how they will translate *Agenda 21* into action at a country level. These strategies are often developed by National Councils for Sustainable Development (NCSDs), a multi-stakeholder participatory body, existing in more than 70 countries. Their status varies from region to region (they are very active in Latin America, moderately so in Asia, limited in Africa) but, where they exist, NCSDs have sometimes played an important role in promoting dialogue and participatory decision-making processes. They have the potential to play a similar facilitating role in developing strategies for sustainable development – although they will need to broaden out from their primary environmental focus to cover the social and economic stakeholders more fully.

An increasing number of countries are developing national visions for sustainable development often supported by the United Nations Development Programme (UNDP) *Capacity 21* programme. National visions bring together different groups of society, including those of different political parties, to agree common development objectives. Examples include Ghana, Tanzania and Thailand (Box 2.11).

Many countries have focused on strategies to reduce poverty. For example, Tanzania developed a Poverty Alleviation Action Plan in 1996, Uganda's Poverty Eradication Action Plan was developed in 1997, and Zambia prepared a poverty alleviation strategy in the late 1990s. The quality of these plans varies. The best are truly cross-sectoral strategies with clearly budgeted priorities. Others are wish lists of social sector investment projects.

---

**Box 2.11** *National visions*

**Ghana Vision 2020** is a policy framework for accelerated growth and sustainable development. It gives a strategic direction for national development over 25 years from 1996 to 2020. Its main goal is national transformation from a poor, low-income country into a prosperous, middle-income country within a generation. This change is expected to be accomplished through a series of medium-term development plans based on the decentralized, participatory planning framework which requires priority-setting at the district level. *Ghana Vision 2020* is the product of an extensive consultation and collaborative effort over some four years involving many groups and individuals from the universities, the public sector, the private sector and civil society, coordinated by the National Development Planning Commission.

The change of government in late 2000 was followed by initial uncertainty about how it would treat *Vision 2020* and the accompanying 2005 policy framework. In May 2001, the new government rejected *Vision 2020* as a framework for formulating economic policies as well as the goal of achieving middle-income status by 2020, reasoning that this goal could not be achieved in the planned time frame, given the major slippages in achieving targets under the First Medium-Term Development Plan (1995–2000). In its place, an alternative vision has been proclaimed – to develop Ghana into a major agro-industrial nation by 2015, propelled by a 'golden age of business'.

The government is currently formulating a new economic policy framework to enable the nation to achieve this new goal and the specifics are yet to be made public. A National Economic Dialogue was held in May 2001, with participation by several stakeholder groups, around six themes: poverty reduction strategy; golden age of business; education, labour market and human resource development; resources for growth; economic policy and financial sector. It was the first national consensus-building exercise for stakeholders to discuss the new government's economic policies, including its approach to poverty alleviation within the context of its new vision for long-term economic growth and the decision to participate in the HIPC programme. Several of the thematic thrusts of the newly evolving economic policy framework cover the same ground as *Vision 2020* (Seth Vordjorgbe, pers. comm.).

**Tanzania's Vision 2025** sets targets to achieve a nation characterized by a high quality of life for all citizens; peace, stability and unity; good governance; a well-educated and learning society; and a diversified economy capable of producing sustainable growth and shared benefits. Implementation is to be through short- and medium-term strategies such as the National Poverty Eradication Strategy, Poverty Reduction Strategy and the Medium-Term Plan.

**Thailand's national vision** was developed over 18 months as part of a participatory process, involving 50,000 people, to prepare the Ninth Economic and Social Development Plan. A draft vision emerged from a first round of consultations in the People's Forum on Development Priorities. This was then subjected to research-based analysis of internal strengths and weaknesses and external opportunities and threats. A revised draft was amended further by the People's Forum, and operational elements related to institutional improvements were added.

---

The Comprehensive Development Framework (CDF) is based on the concept of an holistic approach to development. It was launched by the World Bank in January 1999, to be piloted in a number of countries. A key element of CDF is to encourage a long-term strategic horizon of, say, 15–20 years. It seeks a better balance in policy-making by highlighting the interdependence of all elements of development – social, structural, human, governance, environmental, economic and financial. It emphasizes partnerships among governments, development cooperation agencies, civil society, the private sector and others involved in development. In particular it stresses country ownership of the process, directing the development agenda, with bilateral and multilateral development cooperation agencies each defining their support for their respective plans.

Within this framework, the World Bank and the IMF subsequently launched, in September 1999, a process of Poverty Reduction Strategies for low income countries. Poverty Reduction Strategy Papers (PRSPs) are country-written documents detailing plans for achieving sustained decreases in poverty. Initially required as a basis for access to debt relief in highly indebted poor countries (HIPCs), PRSPs will be required by all International Development Association (IDA) countries as of 1 July 2002. The stated goals of poverty reduction strategies are that they:

> '*should be country-driven, be developed transparently with broad participation of elected institutions, stakeholders including civil society, key development co-operation agencies and regional development banks, and have a clear link with the agreed international development goals – principles that are embedded in the Comprehensive Development Framework*' (World Bank/IMF, 1999).

Guidance for CDF and PRSP explicitly supports building on pre-existing decision-making processes. Governments developing PRSPs and external partners supporting them have taken advantage of this is many cases although, in some, it has taken time for the implications of this approach to be understood.

## Sub-national strategies

In many countries, there are strategic planning frameworks such as district environmental action plans (DEAPs) (eg Box 2.12) and Local Agenda 21s (Box 2.13) at provincial and district levels. Under decentralization, local authorities are assuming devolved responsibility for sustainable development and are required to prepare and implement their development strategies and plans – increasingly through participatory processes, as in Bolivia. Usually, however, the skills and methods to undertake decentralized participatory planning are lacking or weak, the finances to implement plans are inadequate, and, often, local plans have to be passed upwards for harmonization and approval.

As at that national level, visioning is becoming a watchword of strategic planning at sub-national levels. For example, in South Africa under the Municipal Systems Act (2000), negotiated five-year Integrated Development Plans are being developed for all local authorities and District Councils using visioning and priority-setting processes (Box 2.14).

---

**Box 2.12** *The district environmental action plan (DEAP) process in Zimbabwe*

The DEAP process was launched as a pilot exercise in 1995 as a follow-up to the National Conservation Strategy. It is being implemented by the Department of Natural Resources in the Ministry of Environment and Tourism.

The objective is to prepare environmental action plans for all rural districts in Zimbabwe. The first phase concentrated on one ward in each of eight pilot districts – each to include budgeted portfolios for the sustainable development of the natural resource base in the district and one immediately implementable activity to tackle environmental issues identified by villagers in the district.

Following a review in 1997, the initiative was extended in 1999 to cover more wards in the initial pilot districts and a further eight districts – overall there are now two districts per province. The focus of the DEAPs is on poverty alleviation, socio-economic issues and environmental degradation. Activities in each district include:

- developing guidelines for the participatory methodology to be used to engage villagers in identifying environmental problems, setting priorities and initiating action;
- collecting relevant environmental, economic and institutional data in all wards;
- scanning all environment projects/programmes;
- mobilizing technical inputs in developing the plans;
- documenting relevant institutions and expertise, and defining their roles in plan implementation;
- identifying and designing projects/programmes to constitute the main elements of each plan;
- documenting requirements for implementing each plan;
- disseminating each plan among institutions and groups and building consensus on its appropriateness.

The overall programme is overseen by a steering committee of senior officials. Provincial strategy teams are responsible for the training of district, ward and community strategy teams. In each district, a district strategy team is responsible for facilitating the process and reports to the relevant sub-committee of the Rural District Development Committee.

The entry level for activities is now at the ward rather than community level – the latter was judged to have failed and the training in the use of participatory methods introduced in the first phase has ceased.

This initiative is very much on an experimental basis. It is following good principles, but it has yet to be proven on a wide scale. The transactional costs of scaling-up such a comprehensive approach are considerable, especially given the weak capacity of local councils.

*Source:* Munemo (1998)

---

## Local-level strategies

In developing countries, there is a lot of experience of village planning. Nowadays, this planning is being undertaken in a more strategic, participatory and transparent manner. In Tanzania, the HIMA (*Hifadhi Mazingira,* conserve the environment) programme and the *Tanzakesho* (Tanzania tomorrow) programme help wards (three to five villages) to prepare plans through identifying major problems, solutions and sources of required resources. In Nepal, under the Sustainable Community Development Programme,

---

**Box 2.13** *Local Agenda 21s*

Local Agenda 21s can address many weaknesses in local development planning and environmental management – they have increased the willingness of citizens, community organizations and NGOs to 'buy in' to planning and environmental management. They also have some potential to integrate global environmental concerns into local plans. But there are two major limitations of Local Agenda 21s:

1   Their effectiveness depends on accountable, transparent and effective local government – although they can also become a means for promoting these qualities.
2   They have so far given little attention to less obvious environmental issues such as the transfer of environmental costs to other people and other ecosystems, both now and in the future.

The development of Local Agenda 21s has led to considerable innovation in urban areas across the world, including initiatives to encourage city governments to share their experiences. Thousands of urban centres report that they have developed a Local Agenda 21. Many of these have led to practical results and impacts but some may be no more than a document setting out goals or plans of government agencies developed with little consultation – they may simply be conventional plans renamed. Other Local Agenda 21s have been very participatory and contain well-developed goals, yet have foundered because of the limited capacity of city authorities to work in partnership with other groups.

   Several assessments can be found at www.iclei.org. They show that an important challenge for effectiveness has been harmonizing national and local regulations and standards. Unless local actions and regulations are supported within national policy and regulatory frameworks, they cannot be effective. The establishment of a national association of local authorities can help to provide a collective voice.

---

community-based organizations have been trained to develop community plans reflecting shared economic, social and environmental priorities.

A variety of local-level strategies has developed through mechanisms which are largely ignored by central government, but which could provide local pillars for a sustainable development strategy and its supporting coordination system. Some involve traditional assemblies (such as the *khotla* village meetings in Botswana and the Maori *hui* meetings in New Zealand) in which local groups can express concerns and agree actions to build locally appropriate, more sustainable societies.

NGOs often mobilize local energies to combine socio-economic development and environmental conservation at the grassroots level. For example, in Northern Pakistan, the Aga Khan Rural Support Programme is now the leading organization supporting rural development. In Bangladesh, wetland systems have been successfully managed in recent years by NGOs working with the Department of Fisheries. Resource user groups can also play an important role. For example, in Nepal, over the past 40 years, some 9000 forest user groups have assumed responsibility from government for the sustainable management of parcels of national forests and play an important role in sustainable development in remote villages.

---

**Box 2.14** *Integrated Development Plans, South Africa*

In South Africa, the Directorate of Land Development Facilitation in the Department of Land Affairs is responsible for guiding the implementation of the 1995 Development Facilitation Act (DFA). This Act consists of a number of different elements for which the responsibility is shared between all three spheres of government. The most important element is the setting of Land Development Objectives (LDOs) which provided a negotiated five-year development plan for all local authorities and District Councils.

The Municipal Systems Act (MSA), 2000, addresses the new roles of local government in promoting socio-economic development. Local authorities are expected to take on an enabling role as well as undertake service delivery. They are also expected to assume a strategic orientation. Chapter Five of the MSA deals exclusively with integrated development planning, requiring that each municipality must, within a 'prescribed period after the start of its elected term', adopt a 'single, inclusive and strategic plan for the development of the municipality'.

Such a plan must reflect:

- the Council's vision for the long-term development of the municipality, with special emphasis on the most critical development and internal transformation needs;
- an assessment of existing levels of development, and an identification of communities which do not have access to basic municipal services;
- the Council's development priorities and objectives;
- the Council's development strategies, which must be aligned with any national or provincial sectoral plans;
- a spatial development framework, to guide land use management;
- the Council's operational strategies;
- a financial plan;
- key performance indicators and performance targets.

The performance of the municipality will be monitored and evaluated against the objectives set in the Integrated Development Plan (IDP).

Chapter Four of the MSA deals with 'community participation' in local government. It prescribes that municipalities must develop 'a culture of municipal governance that complements formal representative government with a system of participatory government'. Municipalities must 'encourage, and create conditions for, the local community to participate in the affairs of the municipality' – including the drafting of the IDP. Communities must participate in the establishment of a Performance Management System, performance monitoring, the preparation of budgets, and strategic decisions relating to the provision of municipal services. Municipalities must also contribute to 'building the capacity of the local community to enable it to participate in the affairs of the municipality, and of councillors and staff to foster community participation'.

*Source:* Khanya-mrc (2000, 2001)

---

## Some common features of existing strategic planning processes

Many of these planning approaches have been undertaken on the presumption that governments drive and dominate the development process.

Many of the national and sectoral plans overlap in subject and geographical scope, and some have been duplicated in the same country.

Some form of public consultation has been a feature of the development of most conservation strategies, environmental action plans and of some sectoral plans and strategies, but rarely have the results of such consultations been included in any obvious way in the final documents. The preparation of national plans or strategies has frequently involved (foreign) consultants talking to the same small group of people in governments and the public service who, in turn, have had to meet conflicting demands. One notable advance has been the recent work to develop a National Forestry and Conservation Action Plan (NFCAP) in Papua New Guinea. The extensive consultation of all types of NGOs in this process has resulted in a sharing of responsibilities for NFCAP planning and implementation between NGOs and government. It has also stimulated NGOs to form a coordinating organization (Mayers and Peutalo, 1995).

The CDF, PRSP, national visions and local-level planning initiatives encompass a significant number of common principles (see Box 2.15). They also demonstrate the potential for a convergence of approach with the concept of a sustainable development strategy. Often there are several such initiatives on-going in a country. A greater convergence around the principles of strategic planning would ensure complementarity and coherence between national level strategies and ensure that the links between national and local level planning are developed effectively.

### Guidance on strategies for sustainable development

Five years after the Rio Earth Summit (UNCED), the UN General Assembly Special Session review meeting in 1997 revisited the issue of national strategies for sustainable development (NSDS) and set a target date of 2002 for introducing such strategies in all countries. Similarly, the Organisation for Economic Co-operation and Development (OECD) Development Assistance Committee, in its *Shaping the 21st Century* document, set 2005 as a target date for the same goal (OECD-DAC, 1997a), while the OECD's *Principles for Capacity Development in Environment* stresses the importance of strategies providing a framework for donor coordination (OECD-DAC, 1997b).

The Millennium Development Goals include one to 'integrate the principles of sustainable development into country policies and programmes and reverse the loss of environmental resources' (UNGA, 2001, Goal 7, target 9).

Most recently, the Plan of Implementation agreed at the World Summit on Sustainable Development recommits governments to taking action on NSDSs:

> *'States should:*
>
> *(a) Continue to promote coherent and coordinated approaches to institutional frameworks for sustainable development at all national levels, including through, as appropriate, the establishment or strengthening of existing authorities and mechanisms necessary for policy-making, coordination and implementation and enforcement of laws;*

*(b) Take immediate steps to make progress in the formulation and elaboration of national strategies for sustainable development and begin their implementation by 2005. To this end, as appropriate, strategies should be supported through international cooperation, taking into account the special needs of developing countries, in particular the least developed countries. Such strategies, which, where applicable, could be formulated as poverty reduction strategies that integrate economic, social and environmental aspects of sustainable development, should be pursued in accordance with each country's national priorities'* (Paragraph 145, Plan of Implementation, World Summit on Sustainable Development, 4 September 2002).

There is little doubt that those countries with a clear national sustainable development strategy are more likely to secure investment as aid agencies move away from the funding of uncoordinated stand-alone projects and demand that development projects fit into national frameworks. However, despite these international targets, until recently, there has been a lack of clarity on what an NSDS actually is and no internationally agreed definition. But there is a growing view that the focus of an NSDS should be on improving the integration of social and environmental objectives into key economic development processes and should, therefore, be seen as 'a strategic and participatory process of analysis, debate, capacity strengthening, planning and action towards sustainable development' (OECD-DAC, 1999).

This view also holds that an NSDS should not be a completely new planning process to be conducted from the beginning. Rather, it is recognized that in an individual country there will be a range of initiatives that may have been taken in response to commitments entered into at UNCED or as part of commitments to international treaties and conventions, and that these may be regarded in that country, individually or collectively, as the NSDS. Some of these initiatives are government-sponsored, some are led by NGOs and others are multi-stakeholder initiatives. Many are built on earlier or existing processes such as national conservation strategies and national environmental action plans. The challenge is to clarify what initiative(s) are regarded in the country as making up its NSDS and, then, to identify what improvements need to be made to these initiatives, or developed between them such as a coordination system or an umbrella framework, systems for participation and national sustainable development forums.

A useful initiative was undertaken during 2000–2002 by the OECD Development Assistance Committee (DAC) to learn from past and current experiences of developing countries in undertaking strategic planning processes. The first phase of this initiative involved stakeholder dialogues in a number of countries which identified best practice and provided lessons and guidance on how development cooperation agencies can best help developing countries in developing and implementing NSDS. A focus of this work was the development of principles for NSDS (Box 2.15).

A second phase involved the preparation of a resource book on strategies which reviews and analyses a wide range of past and current strategic planning approaches and provides guidance on how to develop, assess and implement

**Box 2.15** *Key principles for strategies for sustainable development*

These are principles around which strategies should be developed, and towards which they should aspire. All principles are important and no order of priority is implied. They encompass a set of desirable processes and outcomes that also allow for local differences. A more specific checklist of criteria could be developed from these principles to suit local conditions, and could then be used for monitoring the progress of a strategy.

*1 People-centred:* an effective strategy requires a people-centred approach, ensuring long-term beneficial impacts on disadvantaged and marginalized groups, such as the poor.

*2 Comprehensive and integrated:* strategies should seek to integrate economic, social and environmental objectives. Where integration cannot be achieved, trade-offs need to be negotiated. The entitlements and possible needs of future generations should be factored into this process.

*3 Country-led and nationally owned:* it is essential that countries take the lead and initiative in developing their own strategies if they are to be enduring. Past strategies have often been driven by external pressure and development agency requirements.

*4 High-level government commitment and influential lead institutions:* these – on a long-term basis – are essential if policy and institutional changes are to be seen through, financial resources are to be committed, and for there to be clear responsibility for implementation.

*5 Effective participation:* broad participation can help in many ways: to open up debate to new ideas and sources of information; expose issues that need to be addressed; enable problems, needs and preferences to be expressed; identify the capabilities required to address them; and make possible a consensus on the actions to be taken that leads to better implementation. Central government must be involved to provide leadership, shape incentive structures and allocate financial resources; but multi-stakeholder processes are also required – involving decentralized authorities, the private sector and civil society, as well as marginalized groups. This requires good communication and information mechanisms with a premium on transparency and accountability.

*6 Based on comprehensive and reliable analysis:* priorities should be established on the basis of a comprehensive analysis of the present situation and of forecasted trends and risks, examining links between local, national and global challenges. The latter include, for example, those resulting from globalization or the impacts of climate change. Such analysis depends on credible and reliable information on changing environmental, social and economic conditions, pressures and responses. Local capacity for analysis and existing information should be fully used, and different perceptions among stakeholders should be reflected.

*7 Link national and local levels:* strategies should be developed in a two-way iterative process within and between national and decentralized levels. The main strategic directions should be set at the central level (here, economic, fiscal and trade policy, legislative changes, international affairs and external relations, etc, are key responsibilities). But detailed planning, implementation and monitoring are better done at a decentralized level, with appropriate transfer of resources and authority.

8 *Consensus on long-term vision:* strategic planning frameworks are more likely to be successful when they have a long-term vision with a clear time frame upon which stakeholders agree. At the same time, they need to include ways of dealing with short- and medium-term necessities and change. The vision needs to have the commitment of all political parties so that an incoming government will not dismiss a particular strategy as representing only the views or policies of its predecessor.

9 *Building on existing processes and strategies:* a strategy for sustainable development should not be designed as a new planning process but instead should enable convergence, complementarity and coherence between different planning frameworks and policies that are already well established. This requires good management to ensure coordination of mechanisms and processes, and to identify and resolve potential conflicts. The roles, responsibilities and relationships between the different key participants in strategy processes must be clarified early on.

10 *Develop and build on existing capacity:* at the outset of a strategy process, an assessment should be undertaken of the political, institutional, human, scientific and financial capacity of potential state, and civil society participants. Where needed, provision should be made to develop the necessary capacity. A strategy should optimize local skills and capacity both within and outside government.

11 *Targeted with clear budgetary priorities:* a strategy for sustainable development must be fully integrated into the budget process to ensure that plans have the financial resources to achieve their objectives (and are not mere 'wish lists'). The formulation of budgets must be informed by a clear identification of priorities. Capacity constraints and time limitations will have an impact on the extent to which the intended outcomes are achieved. Targets need to be challenging – but realistic in relation to these constraints.

12 *Incorporate monitoring, learning and improvement:* monitoring and evaluation needs to be based on clear indicators and built into strategies to steer processes, track progress, distil and capture lessons, and signal when a change of direction is necessary.

*Source:* Adapted from OECD-DAC (2001)

strategies for sustainable development (OECD/UNDP, 2002). It has a particular focus on tried, tested and practicable approaches. Details of this DAC project and a comprehensive set of key literature on NSSD can be found on the website www.nssd.net.

The key points set out in the DAC principles for strategies are winning wider international acceptance. For instance, they are fully reflected as strategy process guidelines in the report of a UN expert forum on national strategies for sustainable development held in November 2001 in Accra in preparation for the 2002 World Summit on Sustainable Development (UN DESA, 2002).

## *A continual learning approach*

For most countries, there is unlikely to be a single national sustainable development strategy. Rather the notion is likely to encompass the sum of the different initiatives being undertaken in an individual country, but with effective coordination, stakeholder participation and a focus on processes which provide for continual learning and improvement. Latest thinking on such an approach, and case studies of developing country experience can be found in Dalal-Clayton et al (2002).

The processes of developing strategies for sustainable development should create an enabling environment, not a strait jacket – particularly for district and local planning. In diverse and complex rural areas, creative and flexible participatory learning approaches will need to be tried and tested, with institutions adapting to local realities – not the reverse. In Chapter 5, we consider some options.

## Sustainable development indicators

Since the 1972 UN Conference on the Human Environment in Stockholm, several attempts have been made to define general indicators of sustainable development (eg Kuik and Verbruggen, 1991; Hammond, 1995; Winograd, undated). These have never been applied successfully at the level of a project or set of activities, as they are frequently too broad to be useful in practice.

Renewed interest followed the 1992 'Earth Summit' in Rio de Janeiro, and the last decade has seen considerable efforts by governments, NGOs and multi-stakeholder groups to develop indicators for sustainable development. The Consultative Group on Sustainable Development Indicators was established in 1996 to promote cooperation and coordination among institutions and individuals; to arrive at a small number of sustainable development indices; and ultimately develop a single, internationally accepted sustainable development index to supplement GDP/GNP and other measures of progress. The International Institute for Sustainable Development (IISD) provides a coordinating secretariat for this work.[4] Its website provides a useful entry point to this subject (www.iisd.org), and provides a global compendium of indicator initiatives (some examples are listed in Box 2.16).

While it is possible to establish such indicators of sustainable development, the availability and reliability of basic data by which to measure them is likely to be a problem in developing countries. This is one reason why there has been a tendency to continue to rely on easily derived indicators such as GDP – which are rather meaningless in terms of sustainable development.

Cost-effective ways for communities themselves to gather the data needed must still be found.

Broad, national-level indicators may have some use in national strategic

---

4 CGSDI Secretariat, International Institute for Sustainable Development, 161 Portage Avenue East, 6th Floor, Winnipeg, Manitoba R3B OY4, Canada. Tel: +1 204 958 7700; fax: +1 204 958 7710; email: phardi@iisd.ca.

**Box 2.16** *Some initiatives on indicators for sustainable development*

## Worldwide

- *UNEP:* Project to measure states of and trends in the environment and guide policy-making towards sustainable development, in implementation of United Nations Environment Programme (UNEP) Environmental Observing and Assessment Strategy. Development of indicators in specific sectors; approaches to aggregation of indicators; use of indicators in state-of-the-environment reporting. Indicator sets issues in 1997, 2000 and 2001 (http://earthwatch.unep.net/indicators/index.html).
- *World Bank:* (a) The Bank's annual *World Development Indicators* (WDI) includes 800 indicators in 75 tables, organized in 6 sections: world view, people, environment, economy, states and markets and global links. The tables cover 148 economies and 15 country groups – with basic indicators for a further 58 economies (www.worldbank.org/data/wdi2000); (b) The Environmental Economics and Indicators Unit has developed indicators of environmentally sustainable development and environmental performance (for WB projects) (www.worldbank.org/environmentaleconomics).
- *UNDP: Human Development Reports* published since 1990, presenting the Human Development Index (HDI) as a measure of human development in individual countries (http://hdr.undp.org).
- *OECD:* programme to develop a core set of (and supporting sectoral) environmental indicators initiated in 1990 (based on policy relevance, analytical soundness and measurability) (www.oecd.org/dac/indicators/index.htm).
- *Dow Jones Sustainability Group:* Indexes (one global, three regional and one country) based on the world's first systematic methodology for identifying leading sustainability-driven companies worldwide (www.sustainability-index.com).
- *UN Division for Sustainable Development (DSD):* Developed working set of 134 sustainable development indicators for decision-makers at national level – collaborative effort building on indicator work in several organizations. Twenty-two countries agreed to voluntarily test the indicators over 3 years (www.un.org/esa/sustdev/isd.htm).
- *World Resources Institute:* Project on highly aggregated, policy-relevant environmental indicators – developed map-based indicators of biodiversity and land use; and indicators of material flows (national, sectoral and company levels) (www.wri.org).
- *Sustainable Measures (*formerly *Hart environmental Data):* Comprehensive website with a database of indicators-related projects and resources to help people and organizations with their indicator research. Specializes in community indicators (www.sustainablemeasures.com/Indicators/index.html).

## National/provincial

- *United States:* Interagency Working Group on Sustainable Development Indicators developed experimental set of 40 indicators to encourage a national dialogue towards developing a set of national indicators (www.sdi.gov/reports.htm).
- *Finland:* Finnish Environment Institute developed sustainable development indicators for use at national level – 20–30 indicators in each of 4 categories: environmental, economic, social, conflict (www.vyh.fi/eng/welcome.html).
- *United Kingdom:* Department of the Environment, Transport and the Regions developed core set of 150 (and 15 headline) sustainable development indicators to

be central to future progress reports – take an economic-social-environmental-resource model while recognizing interactions between them (www.sustainable-development.gov.uk/indicators).

### County, municipal, local area, community-based

*   *Lancashire County Council, UK:* Second Green Audit incorporates 40 sustainability indicators for the county (www.la21net.com).
*   *City of San Jose, California:* The City Policy and Planning Division developed 52 quantifiable indicators of sustainability in 9 categories as a step towards creating a centralized, coordinated environmental data system for performance measurement and public information (www.ci.san-jose.ca.us/esd).
*   *Hamilton-Wentworth, Ontario, Canada:* Sustainability indicators to monitor progress towards goals of city's Vision 2020 – developed through community consultation process using workbooks (www.vision2020.hamilton-went.on.ca).
*   *Sunrift Center for Sustainable Communities, Minnesota, USA:* Developed the Flathead Gauges to identify/quantify key components of sustainability in Flathead County, and to measure trends. Involved public meetings, citizen survey and feedback from individuals and organizations (cdaly@netrix.net).
*   *Sustainable St Louis:* Measure of St Louis project assists citizens to develop and monitor a set of community-defined indicators of sustainable development (Contact: Claire Schosser, PO Box 63348, St Louis, MO 63163, USA; Fax: +1 314 773 1940).
*   *Integrative Strategies Forum, Washington, DC:* The Metro Washington Community Indicators Project is a voluntary initiative, promoting the development and use of community indicators as part of a wider sustainability planning process (jbarber@igc.org).
*   *Sustainable Northern Ireland Programme:* An NGO working with communities and local authorities to promote Local Agenda 21 and sustainable development in Northern Ireland. Helped several councils to develop initial sets of indicators to raise public awareness about sustainability
    (Contact: Michael@snipl.freeserve.co.uk).

*Source:* www.iisd.org

planning, as they shed some light on a country's development priorities and trajectory over a period of time. Nevertheless, these indicators are so broad as to be irrelevant in practice to most applications of rural planning. For these reasons, we do not examine the issue of sustainable development indicators in detail in this study. Rather we take the view that a good planning system (at whatever level – national to local) should have a monitoring system (with indicators) in order to track progress. As a consequence, indicators for rural planning purposes are best developed on a case-by-case basis as part of the planning process itself. Local criteria emerging from this process should be the primary guide to suitable indicators. As in the rural planning process itself, local communities should be involved in identifying locally meaningful indicators.

# PROS AND CONS OF CONVENTIONAL APPROACHES

Land use opportunities and problems stem from both land and people. They involve physical, biological, social and economic issues, so decision-makers need information on each of these aspects. Integrating this diverse information is difficult, and is usually done intuitively, the decision-maker placing greatest reliance on the information that s/he best understands and perceives to be important.

The decision-making process may be represented as a series of 'What if?' questions, the answer to each determining the next question. This means that decision-makers need dynamic interpretations of data on natural resources and the technology by which they are exploited that keep pace with the changing nature of the questions asked of them. The decision-makers' question is not simply 'What is the depth to groundwater?' but a sequence: 'Is there groundwater at reachable depth (if so, what depth) and for how long will there still be groundwater at reachable depths if we install wells every km and pump at 10,000 gallons a day?' Answers have to be area-specific, so that particular areas, people and projects can be identified and, usually, answers have to be quantitative.

Throughout the cycle of a development initiative, from conception to feasibility study, to bidding for funding, to detailed design and implementation, to monitoring and evaluation, the need for information changes. A different mix of political, social, economic and biophysical information and a different level of detail or generalization is needed for sequences of decisions that have to be made at different stages of the cycle. And if the project extends over many years, new needs appear and old data require updating (see Box 2.17).

Ideally, policy-makers and managers need to use the data iteratively in the course of the evolution of projects, planning exercises or development programmes. A static, one-shot survey cannot fulfil this need. At present, poor countries lack the capability for re-interpreting, updating and upgrading their natural resources data base.

## Common limitations of natural resources surveys

### Terms of reference

The most important part of any natural resources survey is a clear statement of purpose. Explicit objectives are needed to avoid, on the one hand, the danger of collecting irrelevant data and, on the other hand, omitting something important. Terms of reference for natural resources surveys have usually failed this test. The objectives of the whole exercise have been vague while the operational procedures have often been laid down in some detail usually copied from elsewhere. Thus, they have imposed standard recording of standard characteristics, standard intensities of survey and standard interpretations.

This criticism applies equally to donor organizations supporting resource surveys for development projects or supporting survey organizations and to indigenous survey organizations that simply follow established, scientifically

---

**Box 2.17** *The accelerated Mahaweli project, Sri Lanka*

Confirmation that there was enough water and enough land for very big hydro-power and irrigation developments in the Mahaweli basin came from reconnaissance land resources surveys (especially Hunting Survey Corporation, 1962) and a good topographic base map at one inch to one mile. An order of magnitude more of analysis (and eighteen volumes instead of one) was needed to produce a *master plan* for basin development (FAO/UNDP, 1969), although little more systematic resources survey was undertaken. Decisions by international donors to fund the elements of the Accelerated Mahaweli Project were taken mainly on political grounds, although the economic case for the hydro-power component alone justified these decisions.

Feasibility studies undertaken to terms of reference laid down by the donors *followed* the decision to fund, which was taken on trust. In the event, these feasibility surveys served as design and implementation surveys of the land development component which went ahead without any more detailed information, other than topographic surveys.

Sophisticated models of land use recommended by the consultants have not been adopted. Neither the administrative and planning capacity nor the natural resources database are adequate. Farmers in the Mahaweli area grow rice, and a lot of water seeps through permeable soils whose distribution or permeability is not known (neither soil permeability, available water capacity nor even soil texture are diagnostic characteristics of the soil classification used in the soil survey).

A new phase of investment with detailed attention to market opportunities, product processing, infrastructure, hydrology, soil permeability and nutrient status, agronomic development and extension will be needed to upgrade the farming system to meet the social and economic challenges of the next few decades. Equivalent effort, again needing better land use, soil, hydrology and farming systems information will be needed to arrest land degradation in the catchment and protection of water supplies which were not seriously considered in the original development.

*Source:* Dent and Goonewardene (1993)

---

respectable methods. Sometimes the procedures are modified but, essentially, no new thinking is involved. As a result, in feasibility surveys, excessive time is spent fulfilling the terms of reference and early data are not available to other team members when they are framing their own concepts. Soils data, in particular, are all too often placed in appendices – collated long after the recommendations have been made – and remain largely unused. In systematic national surveys, the responsible organization often runs itself into the ground with little real coverage to show for many years of effort. So when a question is asked about a particular area, the answer is likely to be 'We have no information for that area'.

In fact, terms of reference and surveys tend to be the product of the same, quite small, group of people. There is an incestuous cycle that breeds overweight surveys that serve the scientific and professional interests of the instigators but are ill-adjusted to the needs of decision-makers. To give three examples:

1   Ive et al (1985) noted that in the South Coast Project in New South Wales (Austin and Cocks, 1978), 150 data items were collected against 5300 mapping units, but only 20 data items were used in the subsequent land allocation phase.

2   For a resource inventory of the Gambia (Dunsmore et al, 1976), LRDC expended more than 20 man-years of effort, the greatest component of which was in establishing soil series characterized by detailed laboratory data, only to publish the soil information as 1:100,000 maps of soil associations that could have been completed in a tenth of the time. Subsequently, consultants to the Gambia Barrage Project (Coode and Partners, 1979) found key data lacking on the soils of the tidal flood plan (which had resisted soil series characterization), and no information on contours, river discharge, land use or mangrove timber resources. A rapid appraisal, which discovered 13,000 ha of potential acid sulphate soils in the project area, and dynamic modelling based on approximate data, killed the project in three man-months. These results were then confirmed by a further three man-years of conventional survey and laboratory analysis (Thomas et al, 1979). LRDC surveys in The Gambia, classics of their kind, simultaneously achieved overkill of superfluous data and missed crucial information needed for development. All the original soil data were, subsequently, thrown away.

3   Dent and Goonewardene (1993) describe the collection of exhaustive data on market opportunities, farm economics and regional socio-economic evaluation by an FAO team in Sri Lanka. More than 6 metres of shelf space, some documents costing more than £1000 a page at today's prices, assembled as the basis for a land rehabilitation and settlement project, still lies unused by the project management. The natural resources data, though collected and analysed by state-of-the-art methods, is incomprehensible to the decision-makers; the planning recommendations are too complex; the recommended farming system is unpopular; and the market economics data are now out of date.

## Comprehension

Natural resources data are under-utilized because land users, planners and other professionals, let alone policy-makers, do not appreciate their utility. There are several reasons for this. The two most commonly complained about, though not necessarily the most fundamental, are:

1   First, they are simply not understood by anyone except the specialists who produce them. Jargon and the intimidating welter of data are obvious reasons for this. Natural resources reports are, quite clearly, written for the benefit of natural resources specialists and not for anyone else. Reports should be addressed to the users and presented in a way that they can clearly understand, which means that the users should be known in the first place.

2   Secondly, the information does not tie in with the experience of the potential users. Apart from the language, the level of detail or generalization

of the information is not the same, and the resource mapping units may not be recognized by the decision-maker. Information must coincide with the effective management unit, generalized at national level but detailed and precise at field level, if its applications are to be recognized.

## Usefulness

Natural resources information of all kinds is very interesting to natural resources professionals. It helps them build up their model of the world. But much is irrelevant to the kinds of decisions actually being taken about land use. Box 2.17 draws attention to the use in irrigation projects in Sri Lanka of a pedological soil classification that ignores permeability and available water capacity – the two most important soil characteristics for irrigation. Perhaps the most common use of inappropriate data is the use of small-scale survey data for detailed planning because no large-scale data are available. For most developing countries, the best soil map is still the FAO 1:5 million soil map of the world completed 30 years ago!

That there is a problem of comprehension is well understood. In response, survey interpretations have been provided both to simplify the data and to integrate them. Examples include the Storie Index, land capability classification and other approaches discussed earlier in this chapter. Each of the well-known interpretations was originally designed for a specific, practical purpose which it fulfils very well. However, it has since been adopted as a stock interpretation, offered off-the-peg by survey organizations or demanded by clients simply because it is less intimidating than, for example, a sheaf of soil maps.

In truth, the usefulness of an off-the-peg interpretation is extremely limited and this is especially so in poor countries that are already densely settled and farmed to the limits, or beyond the limits, of the present capacity of the land. We are not starting with a clean sheet. Decisions already taken and acted upon severely limit our room for manoeuvre and it is not helpful to a subsistence farmer to tell her that her land is suitability class S3 or even class S2 for millet! If all that comes out at the end of an exhaustive natural resources survey is a ranking of S1 to S3 or N, essentially on the basis of slope angle, the cost-effective procedure would have been simply to measure the slopes.

Relatively few mature systems of resource survey and evaluation have evolved. Where they have, they have arisen within stable, well-resourced organizations – not, by definition, in developing countries. There appears to be little or no consultation between land resources professionals and any of the supposed beneficiaries of development or supposed users of information. Natural resources professionals, rather than users, drive the methods used. Henry Ford had an expression for it: 'The customer can have any colour he likes, so long as it's black.' As a consequence, natural resources surveys and land evaluation are usually addressing yesterday's problems.

The syndrome is exaggerated in developing countries where scientific and technical staff are few and very inadequately resourced. Their natural resources professionals do not have the benefit of frequent exposure to a wide range of approaches and methods. They work in an institutional environment where skills and information are hoarded rather than shared, and they find it safer to continue the status quo or defer to outside consultants than to strike an independent course. Short-term consultants are also constrained by inappropriate terms of reference and an unwillingness on the part of their clients to allow any change of these, suspecting attempts to cut corners for some commercial advantage.

## Inappropriate planning methods and inappropriate data: a failure of institutions

Development planners have to integrate several discrete sets of information:

- Development is not sustainable if it will degrade or even destroy the natural resources on which it depends. Therefore, information is needed about the condition and trends of natural resources. Information about their present status is only the starting point. We also need to know about the interactions between the resources and their use under each management option being considered.
- Unless the development under consideration meets the need for production and profit, it is unlikely to be adopted. Data on demand, production, markets, costs and returns are essential. The difficulty of forecasting future costs and prices is evaded to some extent by discounting but this gives a distorted, essentially short-term perspective. In particular, private profitability may not coincide with the needs of the wider community or the necessity of conserving resources for the future.
- Any development brings an element of risk, which must be assessed and judged to be acceptable by the parties concerned if a plan is to be implemented. Risks lie in both biophysical factors like the reliability of rainfall, and in the uncertainties of markets and availability of essential inputs.
- Proposals must marry with the social structure and mores of the place. In particular, unless they are accepted as equitable they will not command general support. Rights of land tenure and water use are often key issues that are difficult and dangerous to tackle but which are often constraints upon the sustainable use of resources.
- Information is needed about the motivation of the people, their perceptions and aspirations.

No one set of data is all important. The failure of so many development plans that have emphasized just one physical, economic or social component demonstrates that little benefit is likely to be achieved unless all relevant aspects are considered. This is a big task. It cannot be achieved without better planning procedures and a revolution in the attitude and training of planners and decision-makers.

Failures of land use planning have been as much a failure in working with people as a failure of natural resources data. Land use planning has been a centralized and top-down activity. This is well illustrated by the Zambian *Land use planning guide* (Zambia Department of Agriculture, 1977):

**Table 2.4** *Pros and cons of conventional natural resource surveys and rural planning in developing countries*

|  | Pros | Cons |
|---|---|---|
| Natural resource surveys | Provide vital baseline information | Policy and development decisions are usually taken on political and economic grounds. Natural resources (NR) information plays a minor role |
|  | Methods are well proven and provide reliable data | Surveys and recording networks demand years of effort by skilled staff. Standard data are often not directly relevant for planning or decision-making. Standard data and interpretations are often used for purposes for which they were not designed |
|  | Consultants are able to interpret natural resources data for a variety of purposes, including forecasts and assessment of development options | Planners and decision-makers are confounded by the jargon and intimidated by the welter of details. Survey data are usually static, and lacking interpretation of interactions between resources and their use |
|  |  | There is little consultation between end-users and providers of NR information |
|  |  | NR professionals are few and usually inadequately resourced |
| Planning | Plans provide a necessary framework for development and management of natural resources; guide investment to areas where it will have greatest benefit; and coordinate local and government initiatives | Planning has been a centralized, technocratic activity. Participation by local people has been weak at best and, usually, lacking |
|  |  | Plans have been descriptive, lacking analysis and vision |
|  |  | Planning has been predominantly sectoral and coordination between institutions absent or very weak |
|  |  | Linkages between district authorities and both local communities and higher-level authorities are weak |
|  |  | Planning departments are overloaded, especially since decentralization, and chronically under-resourced. There is a chronic lack of professional capacity |

*'The aim of [catchment conservation] planning would be to* direct the people *to cultivate suitable land, to use the best methods applicable to the area and to make sure that* controls to land use *are implemented in both the mechanical and cultural spheres.'* (our emphasis)

This approach is flawed because the methods advocated by the Department of Agriculture (or any other department) have not been the best and most appropriate in the eyes of local farmers – and thus official control of land use is scarcely possible except by coercion.

The loads that governments impose on themselves in attempting to plan, implement and administer land use soon exceed their administrative and logistic capabilities, and outstrip both the abilities of their professionals to supply natural resources information and their own capability of using it. Governments must learn the limits of their own capacity. Yet they hanker after the tried and tested and failed procedures of physical planning – in which experts prepare maps that indicate in considerable detail how land should be used. The supposed beneficiaries of development have little opportunity to articulate their needs in terms of development, technology or information. Nor do they have the opportunity to contribute their own local knowledge. Box 2.8 highlights some principles that might bring planning and people closer together.

Table 2.4 summarizes the pros and cons of the conventional survey and planning approach. Conventional surveys are often very good at providing basic data, and these data are potentially of great value. But the shortcomings of the approach are by now all too obvious and radical changes are called for. Possibly, the same holds for the elliptical contributions of economists and other social scientists.

In the next chapter, we examine the experience of more participatory approaches to local planning in rural areas. In Chapter 4, we discuss principles for collaboration and, in the final chapter, chart some ways forward for the planning process and its supporting information services.

# 3

# Approaches to participation in planning

## THE NEED FOR PARTICIPATION

Many governments have taken it upon themselves to manage the land. In so doing, they have spawned top-down systems of control that are inefficient, vulnerable to corruption, and take responsibility away from local people – excluding or marginalizing many groups. Half a century of professional development planning has demonstrated the shortcomings of the top-down approach. Plans drawn up by outsiders, with little or no reference to the priorities of the people who have to implement them, are not implemented – at least not in anything like the shape envisaged by their architects.

Nearly always, the decisions about land use that really matter are still those taken by the actual users of the land. In spite of the planners, some degree of local management for local ends continues and people revert to the old ways once control or financial incentives are withdrawn. There are innumerable instances of external initiatives in soil conservation, grazing control and conservation of forests and wildlife that, if not undermined during implementation, have been discontinued once external funding ceased. For example, an evaluation of 25 projects sponsored by the World Bank recorded that 13 had been abandoned a few years after financial assistance ended, and concluded that the main causes of failure were lack of participation by the local communities and lack of attention to building local organizations (Zazueta, 1995). On the other hand, there are many local examples of successful development based on participation of the local community (see Boxes 3.2, 3.6, 3.10 and 3.12). There are also many instances of the enthusiastic and widespread adoption of new practices that yield immediate benefits to the land users and, at the same time, fit easily into the existing system of management – for instance the spread of minimum tillage farming in southern Brazil (Tengberg et al, 1997).

It has become clear that outsiders cannot necessarily identify local priorities nor understand how best to meet them. So where development is led at national and international level, with specialized agencies planning and supporting development by and for others, the means have now to be created to rebuild broadly based participation in policy-making, planning and management. This is

---

**Box 3.1** *Top-down and bottom-up planning*

*Top-down planning* describes the conventional procedure of systematic surveys, on the basis of which plans are devised centrally and worked out in detail by professional staff to meet goals that are, also, decided centrally. Implementation is, again, typically the responsibility of line ministries or other government agencies. Examples include sectoral plans (forestry, fisheries, agriculture departmental plans); watershed management planning; town and country planning; individual land development projects (eg Accelerated Mahaweli Project in Sri Lanka, The Tanganyika Groundnut Scheme, the Gezira irrigation project in Sudan).

*Bottom-up planning* describes planning that is initiated locally and proceeds through the active participation of the community. The pooled experience and local knowledge of the land users and local technical staff are mobilized to identify development priorities, draw up plans and implement them. The advantages are:

- a strengthening of the community's sense of responsibility and confidence in land resource management, and strengthening local institutions to take on further responsibilities;
- the building of popular awareness of land use problems and opportunities;
- enthusiasm for a plan which has a broad base of ownership;
- the full use of local knowledge and skills, and close attention to local goals and local constraints – whether related to natural resources or socio-economic issues;
- better information fed upwards to higher levels of planning, educating upper levels of management to the realities of the situation on the ground.

The disadvantages are that:

- limited technical knowledge at local level means that technical agencies need a big investment in time and manpower in widely scattered places;
- difficulties occur in integrating local plans within a wider framework;
- local efforts may collapse through lack of higher-level support (or even obstruction);
- the local community's interests may not be well represented, that is village leaders or businessmen may dominate decision-making and local planning;
- local interests are not necessarily the same as national interests.

*Source:* Dalal-Clayton and Dent (1993)

---

essential if the process of development is to be made acceptable to society as a whole (Box 3.1).

The World Bank's (1994) Learning Group on Participatory Development has defined participatory development as:

> *'A process through which stakeholders influence and share control over development initiatives, and the decisions and resources which affect them'.*

It is a measure of the gulf that has opened up between development planning and ordinary community decision-making that such a high-powered group should be needed to define and explain the concept of participation for the benefit of development agencies.

---

**Box 3.2** *Planning by the people for the people in Central and South America*

Indigenous communities in Central America and the Andean region of South America have developed quite effective ways to identify, plan, and carry out activities that meet their collective needs. Community issues are discussed and decided in organized meetings, which are run by elected community leaders and attended by representatives from all households. It is expected that all participants provide input in the discussions, where options are assessed and decisions for action taken. Decisions are made by consensus – or near-consensus – and are binding for all. Culturally sanctioned means of carrying out plans include labour pooling (known as *minga* in the Andes and *tequio* in Oaxaca, Mexico) and cash or in-kind contributions by each household in the community. Enforcement takes the form of social recognition for households that consistently fulfil their duty, or ostracism, fines or incarceration for those who do not contribute their labour to the community's well-being.

*Sources:* Aguirre Beltrán (1962) and Wolf (1957), cited in Zazueta (1995)

---

## PERCEPTIONS OF PARTICIPATION

Participation is nothing less than the fabric of social life. People have always participated in the development of their own livelihood strategies and cultures. Whether through formal or informal organizations, autocratic or democratic means, a variety of structures and procedures has evolved to define and address collective needs, to resolve conflicts, to make plans and take the steps necessary to implement them, see, for example, Box 3.2.

Increased stakeholder participation is now being demanded by international donor organizations, many NGOs and, more recently, government departments. In Namibia, the Ministry of Environment and Tourism sets out to promote more participatory planning in the introduction to its policy document on land use planning:

> *'The Ministry of Environment and Tourism supports rational, sustainable integrated planning of land use in all environments throughout Namibia according to the sound ecological principles contained in Article 95(1) of the National Constitution. To facilitate appropriate land-use planning and subsequent land use, this Ministry supports the process of consultation with appropriate institutions to ensure that local communities are involved in all decision-making processes and to ensure that they enjoy maximum sustainable benefit from the land and natural resources with which they are associated and upon which they depend'* (MET, 1994).

Words are one thing, actions another. *People's participation* and *popular participation*, like sustainability, are now mantras of the development agencies. They appear in the public statements and stances of even those agencies that have little to do with people or participation. Adnan et al (1992) observe:

*'The meaning of the phrase has become even more elusive after its professed adoption by the most unexpected quarters. It is often difficult to understand whether those talking about people's participation mean the same thing or simply use the phrase as a kind of magical incantation.'*

Rahnema (1992) asserts that:

*'People are dragged into participating in operations of no interest to them, in the very name of participation.'*

This has created several paradoxes. The term *participation* has been used, on the one hand, to justify the extension of control by the state and, on the other hand, to build local capacity and self-reliance. It has been used to justify external decisions as well as to devolve power and decision-making away from external agencies. It has been used for data acquisition by experts and for more interactive analysis. The varied perceptions of participation are illustrated by the opinions voiced during the development of the Bangladesh Flood Action Plan (Box 3.3).

---

**Box 3.3** *Some perceptions of participation in the Bangladesh Flood Action Plan*

**Villagers**

- 'Participation is about doing something for everyone's benefit.' (A villager in Gaibandha)
- 'Oh yes, the foreigners were here one day, last month. But they only went to the school and spoke in English. We are not educated. We could not understand.' (A poor peasant)

**Government officials**

- 'Yes, we're doing people's participation. We have had people working in Food for Works programmes since the seventies.' (Top official in Bangladesh Water Development Board)
- 'Your idea regarding women's participation is not correct for the overall national interest.'

**Foreign consultants**

- 'Another idea from the social scientists. Only slogans! First, "poverty alleviation". Then "women" and "environment". Now "people's participation"! It's just a new fad!' (Engineer)
- 'You have to consult my socio-economist, not me. I have no time for this participation. I'm working 12 hours every day on the project.' (Flood Action Plan Team Leader).

*Source:* Adnan et al (1992)

---

The many ways that development organizations interpret and use the term *participation* can be resolved into seven clear types that range from manipulative and passive participation, where people are told what is to happen and act out predetermined roles; to the stage where communities take initiatives on their own which may or may not, challenge the existing distribution of power (Table 3.1). Simply encouraging local people to sell their labour in return for food, cash or materials only creates dependencies. It may give the impression that local people support an externally driven initiative but the impacts rarely persist once the project ceases (Bunch, 1983; Reij, 1988; Pretty and Shah, 1994; Kerr, 1994; Pretty, 1995) and yet development programmes continue to justify subsidies and incentives on the grounds that they bring quick results, that they can win over more people, or that they provide a mechanism for disbursing food to poor people. As little effort is made to build local skills, interests and capacity, local people have no stake in maintaining new practices once the flow of incentives stops.

A study of 230 rural development institutions employing some 30,000 staff in 41 countries in Africa found that, for local people, participation was most likely to mean simply having discussions or providing information to external agencies (Guijt, 1991). Government and non-government agencies rarely permitted local groups to work alone, some even acting without any local involvement. Even where external agencies did permit some joint decisions, they usually controlled all the funding. Another study of 121 rural water supply projects in 49 countries of Africa, Asia and Latin America found that participation was the most significant factor contributing to project effectiveness (Narayan, 1993). Most of the projects referred to community participation or made it a specific project component, but only one in five scored high on interactive participation. The best results were achieved when people were involved in decision-making through all stages of the project from design to maintenance. If they were involved only in consultations, then the results were much poorer.

The terms *consultation* and *participation* are frequently used interchangeably, but consultation is only one form of participation along a spectrum, as illustrated in Table 3.1. Rahnema (1992) concludes that groups that have no stake in defining the balance between economic, social and environmental goals, and between the present and the future, are marginalized by passive, consultative and incentive-driven forms of participation. The 'superficial and fragmented achievements have no lasting impact on people's lives'. If the objective of development planning is to achieve sustainable development – social, economic and environmental – then none of these types of participation will suffice.

In short, participation has been used more as a means for information or communication than for shared decision-making. In industrialized countries, government agencies seem to exhibit what Walker and Daniels (1997, cited in Dubois, 1998b) call the '3 I Model': inform (the public), invite (comments) and ignore (opinions). In developing countries, wherever natural resources are key to livelihoods, and power differences are strong, participation is often limited to community participation, thereby limiting the opportunity to influence any initiatives and preventing any questioning of existing power structures.

**Table 3.1** *Types of participation in local-level development*

| Type | | Characteristics |
|------|---|---|
| 1 | Manipulative participation | Participation is a sham |
| 2 | Passive participation | People participate by being told what has been decided or has already happened. Information shared belongs only to external professionals |
| 3 | Participation by consultation | People participate by answering questions. No share in decision-making is conceded and professionals are under no obligation to take on board people's views |
| 4 | Participation for material incentives | People participate in return for food, cash or other material incentives. They have no stake in prolonging practices when the incentives end |
| 5 | Functional participation | Participation is seen by external agencies as a means to achieve project goals, especially reduced costs. People may participate by forming groups to meet predetermined project objectives |
| 6 | Interactive participation | People participate in joint analysis, which leads to action plans and the formation or strengthening of local groups or institutions that determine how available resources are used. Learning methods are used to seek multiple viewpoints |
| 7 | Self-mobilization | People take initiatives independently of external institutions. They develop contacts with external institutions for resources and technical advice but retain control over how resources are used |

*Source:* Adapted from Pretty (1997)

It might be argued, as does Degnbol (1996), that government commitment towards a bottom-up paradigm of development has been conducted without any meaningful reforms of the power relations between government and local communities. Participation has been seen as a means to ensure the more efficient implementation of preconceived plans, often through existing government structures. Degnbol points out that it is, possibly, naive to expect governments to redefine their roles and that genuine participation will come about only with the emergence of a strong and representative civil society.

All the evidence points to the need to seek all stakeholders' ideas and knowledge and to give the people on the ground the power to make decisions independently of external agencies. The dilemma of the authorities is that they both need and fear people's participation. They need the agreement and support of diverse groups of people – development is not sustainable otherwise – but they fear that their greater involvement is less controllable, less predictable and likely to slow down the planning process. All true: but if this fear allows only stage-managed participation, then distrust and greater alienation are the most likely outcomes.

**Table 3.2** *Levels of participation in policy processes and planning*

| Level | |
|---|---|
| 1 | Participants listening only – eg receiving information from a government public relations campaign or open database |
| 2 | Participants listening and giving information – eg through public inquiries, media activities, hotlines |
| 3 | Participants being consulted – eg through working groups and meetings held to discuss policy |
| 4 | Participation in analysis and agenda-setting – eg through multi-stakeholder groups, round tables and commissions |
| 5 | Participation in reaching consensus on the main strategy elements – eg through national round tables, parliamentary/select committees, and conflict mediation |
| 6 | Participants involved in decision-making on the policy, strategy or its components |

*Source:* Bass et al (1995)

This dilemma is all the more acute in the field of policy-making. Table 3.2 lists six levels of participation revealed by an analysis of policy processes, which closely parallel the typology in Table 3.1. The greatest degree of public participation achieved in policy-making and planning is in reaching consensus on the elements of a strategy (level 5 in Table 3.2). Fundamental decision-making on national policies and strategies (level 6) remains the prerogative of the national decision-making process, democratic or otherwise.

One of a number of methods that has been used to promote participation in policy-making is the *future search conference*, a multi-stakeholder forum introduced by the Australian systems thinker Fred Emery (Emery and Emery, 1978). It has been used throughout the world for a wide range of other purposes, including helping develop a nature tourism strategy in the Windward Islands (Box 3.4), Pakistan's National Conservation Strategy and developing Colombia's energy sector policy. As described by Baburoglu and Garr (1992):

> *'The conference [usually 35-40 participants, 2-3 days] uses a systematic process in which groups design the future they want and strategies for achieving it. The "search" is for an achievable future. This may be a future that is more desirable than the one that is likely to unfold if no action is taken, or a future that is totally unexpected. Designing a future collectively unleashes a creative way of producing organisational philosophy, mission, goals and objectives enriched by shared values and beliefs of the participants. This process is especially useful in times of social, economic and technological turbulence [characterised by unexpected changes, uncertainty, unintended consequences and complexity].'*

---

**Box 3.4** *Search conferences and nature tourism strategies in the Windward Islands*

Search conferences were held in four Windward Island countries in the eastern Caribbean during 1991/92 as part of a process to develop nature tourism strategies. The stimulus was the threat to banana exports in the face of an impending change in trade relations with the UK in 1992 that was expected to lead to increased economic dependence on tourism. Limited potential for expanding traditional tourism resulted in a growing interest in nature tourism (also called eco-tourism).

The search conference process was initiated by senior officials in government agencies with a direct interest in tourism (planning/economic development, tourism and forestry). The Canadian *Adapting By Learning* group acted as catalysts. In each country, key stakeholders were brought together by the government body acting as the lead agency – either tourism or planning. Stakeholders included: government agencies (economic development, planning, tourism, agriculture, fisheries, finance, forestry), environmental and heritage groups, community organizations, women's and youth groups, farmers' cooperatives, and private business. An initial task was to form National Advisory Groups to direct the process.

The search conference allowed the political implications of nature tourism to be addressed by the full range of interests involved: eg environmentalists examining the validity of nature tourism as an economic development strategy; hotel associations incorporating environmental conservation into tourism strategies; and finance officials working with agricultural ministry personnel to support small businesses in the provision of local produce for tourist consumption.

The objectives of the conferences were to:

- develop comprehensive national perspectives on nature tourism;
- connect this alternative approach to existing tourism initiatives;
- examine the potential of an integrated nature tourism strategy as a basis for future economic development that is environmentally sustainable;
- discuss the planning, design and management needs of such an approach;
- advise on ways in which the search conference initiative could assist in creating an integrated and on-going planning capacity, both nationally and regionally.

Each conference involved alternating plenary and small group sessions with presentations (by local participants with special skills and experience) to provide a basis for discussions. The small groups generated issues and concerns which were reported in plenary. Key issues were selected to focus subsequent group sessions during which constraints and opportunities were identified. Ideas and concerns were then integrated into a set of recommendations for action, and submitted to the National Advisory Groups to be carried forward into further planning and implementation.

*Source:* Franklin and Morley (1992)

---

# Horizontal and vertical participation

In their study of participation in strategies for sustainable development, Bass et al (1995) distinguish between *horizontal* participation and *vertical* participation. *Horizontal participation* refers to the interactions needed to ensure that issues are dealt with across sectoral interest groups, ministries and communities in different parts of the country. *Vertical participation* is required to deal with issues throughout the hierarchy of decision-making from national to local levels, or from leaders to marginalized groups (see Figure 3.1). The deeper the vertical participation within a given institution or nation, the better the understanding and support for the strategy is likely to be.

In respect of recent national conservation strategies, any participation at national level has been restricted within government and academic contributions. Much multi-disciplinary analysis has been undertaken; there have been improvements to national-level government institutions; some regulatory instruments have been introduced; and policies have been changed, at least on paper. However, there appear to be many blocks to implementation of these policies, for example lack of money, lack of local institutional capacity, or lack of commitment. Across the whole field of development planning, even when all views have been sought and consensus achieved, it may remain difficult for politicians to make an honest response and for establishments to change their ways.

In contrast to the national level, there is ample evidence of interactive participation of communities and sectoral interests at local levels – resulting sometimes in impressive work on the ground, with the generation of much local information and some localized institutional change. Particular progress has been made in:

- joint community/business/local government initiatives in urban or peri-urban areas, often facilitated by local governments and NGOs, such as Groundwork UK, Local Agenda 21s in Australian and UK local authorities;
- buffer zones (economic support zones) around national parks, with joint government/community management; there are many well-documented examples, eg in India, Nepal and Zimbabwe (IIED, 1994a) and several Man and Biosphere Reserves worldwide;
- rural development projects based upon social organization and/or environmental protection, often at river catchment level, again facilitated or managed by NGOs.

## PARTICIPATORY LEARNING AND ACTION

The natural sciences have developed a wide range of objective methods to gather and analyse data but the situation is different when it comes to management information, social issues and determining opinions: you can't stick an auger into a farmer to extract information. Social scientists have usually used *extractive* techniques of their own, such as questionnaires, because large numbers of people can be surveyed and statistics can be applied to determine the reliability of the results. However, these methods are weak at revealing local complexities: many of the contextual grounds for understanding the data are systematically removed or ignored, there is a tacit assumption that the correspondent and researcher hold the same values, and cultural divisions affect the types of response. Multiple perspectives – so essential for knowledge of the

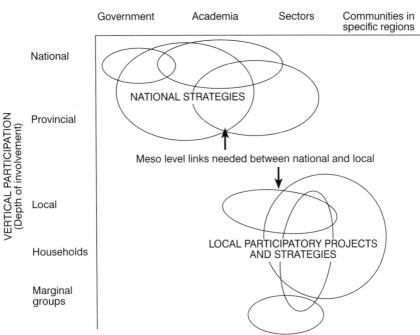

*Source:* Bass et al (1995)

**Figure 3.1** *National and local participation experience*

land and land use – are lost. Gill (1993) has captured a real problem with interview and questionnaire approaches:

> *'The stranger then produces a little board and, clipped to it, a wad of paper covered in what to the respondent are unintelligible hieroglyphics. He then proceeds to ask questions and write down answers – more hieroglyphics. The respondent has no idea of what is being written down, whether his or her words have been understood or interpreted correctly... The interview complete, the enumerator departs and is probably never seen again.'*

There is an alternative that has now won acceptance and credibility. In the 1980s and 1990s, there has been a blossoming of *participatory approaches* – unfortunately accompanied by a babel of acronyms. Some focus on problem diagnosis, eg AEA (Agro-ecosystems analysis), DRR (Diagnostico rural rapido), RRA (Rapid rural appraisal), MARP (*Methode Acceleré de Recherche Participative*); others are oriented toward community empowerment, eg PAR (Participatory action research) and TFD (Theatre for development); some facilitate on-farm or user-led research, eg FPR (Farmer participatory research); others are designed simply to get professionals in the field listening to resource users, eg SB (*Samuhik*

---

**Box 3.5** *Principles of participatory learning and action*

The term *participatory learning and action* (PLA) is now used to encompass a suite of techniques for diagnostic analysis, planning, implementing and evaluating development activities. Principles include:

- *Cumulative learning by all the participants.* Interaction is fundamental to these approaches and the visual emphasis enables all people to take part on an equal basis.
- *Seek diversity* rather than attempt to characterize complexity in terms of average values. Different individuals and groups make different evaluations of situations, which lead to different actions. All views of activity or purpose are laden with interpretation, bias and prejudice. Therefore, there are many possible descriptions of any activity.
- *Group learning.* The complexity of the world will be revealed only through group inquiry and interaction – which requires a mix of investigators from different disciplines, from different sectors, outsiders (professionals) and insiders (local people).
- *Context-specific.* The approaches are flexible enough to be adapted to suit each new set of conditions and actors, so there are many variants.
- *Facilitating role of experts.* The goal is to bring about changes that the stakeholders regard as improvements. The role of the expert is to help people in their particular situation carry out their own study and make their own plans.
- *Sustained action.* The learning process leads to debate about change, and debate changes the perceptions of the actors and their readiness to contemplate action. Action is agreed, so implementable changes will represent an accommodation between different views. The debate and/or analysis both defines changes that would bring about improvement and seeks to motivate people to take action to bring about those changes. This action includes strengthening local institutions, so increasing the capacity of people to take the initiative.

*Source:* Pretty (1994, 1995)

---

*Brahman* – joint trek). Some have been developed in the health context, eg RAP (Resource assessment procedure); some for watershed development, eg PALM (Participatory analysis and learning methods); some in government extension agencies, some in NGOs. The diversity of names, derivations and applications is a sign of strength, because each variation is, to some extent, dependent on its local context. However, they are underpinned by some common principles (Box 3.5), the principal among which is the new learning path that needs to be followed.

Participatory learning and action is the antithesis of teaching and technology transfer, both of which imply transfer of information from one who knows to one who does not know. Its assumptions are completely different from those of conventional surveys, and have grown more distinct as the techniques have evolved. For example, early work in farming systems analysis and rapid rural appraisal was essentially extractive. Researchers collected data and took it away for analysis. There has been a significant shift toward investigation and analysis by local people who share their knowledge and insights with outsiders. Methods

**Table 3.3** *Techniques of participatory learning*

| Group and team interaction | Sampling | Dialogue | Visualization and drawing |
|---|---|---|---|
| • Team contracts | • Transect walks | • Semi-structured interviewing | • Mapping and modelling |
| • Team reviews and discussions | • Wealth ranking and well-being ranking | • Direct observation | • Social maps and wealth rankings |
| • Interview guides and checklists | • Social maps | • Focus groups | • Transects |
| • Rapid report writing | • Interview maps | • Key informants | • Mobility maps |
| • Energisers/ activators | | • Ethnohistories and biographies | • Seasonal calendars |
| • Work sharing (taking part in local activities) | | • Oral histories | • Daily routines and activity profiles |
| • Villager and shared presentations | | • Local stories, portraits and case studies | • Historical profiles |
| • Process notes and personal diaries | | | • Trend analyses and time lines |
| | | | • Matrix scoring |
| | | | • Preference or pairwise ranking |
| | | | • Venn diagrams |
| | | | • Network diagrams |
| | | | • Systems diagrams |
| | | | • Flow diagrams |
| | | | • Pie diagrams |

such as participatory mapping, analysis of air photos, matrix scoring and ranking, flow and linkage diagrams and seasonal analysis are not just means for local people to inform outsiders. Rather, they are methods for local people to undertake their own research (Chambers, 1992). Local people using these methods have shown a greater capacity to observe, create concepts and undertake analyses than most outsiders had expected and are, also, proving to be good teachers.

The techniques of participatory learning fall into four groups (Table 3.3): group and team interaction, sampling, dialogue, visualization and drawing. One of the strengths of participatory inquiry has been the emphasis on pictorial techniques. By creating and discussing a diagram, model or map (see, for example, Figure 3.2), all who are present – both insider and outsider – can see, point to, discuss and refine the picture, sharing in its creation and analysis. Non-literates are not excluded; everyone who can see has visual literacy that allows them to participate actively – although, admittedly, not everyone may be able to speak up in such gatherings.

Early approaches (notably rapid rural appraisal) were, sometimes, quick-and-dirty. As the discipline has evolved, the emphasis has moved from exploitation of local people's labour or knowledge (to push through projects or facilitate research) towards sharing, with contributions from both sides and

*Source:* Denniston (1995)

**Figure 3.2** *Portion from a hand-drawn land use map made by an indigenous surveyor and villagers of the Marwa Sub-Region, Panama*

patient iteration (Box 3.6). This avoids some of the biases of rapid rural appraisal: spatial (eg along roads), personal (led by leaders, entrepreneurs, professionals, English-speakers, males, living), often undertaken only in the dry season (when access is easiest), carried out with politeness/timidity (eg outsiders not shown the worst conditions and will not ask searching questions).

It is not simply the techniques themselves, but the combination and sequence in which they are used, that makes PLA invaluable for understanding the myriad perspectives of natural resource use at the local level. For example:

• Social mapping and well-being ranking can identify diverse socio-economic groups within a community and facilitate an understanding of how wealth and social aspects affect people's dependence on resources.
• Seasonal calendars and time lines show how the use and importance of natural resources varies over the year and over a longer time.

---

**Box 3.6** *RRA and PRA compared*

Among the various approaches of participatory learning and action, perhaps the best known are rapid rural appraisal (RRA) which emerged in the late 1970s and evolved a decade later into participatory rural appraisal (PRA). RRA developed as a response to a growing awareness that conventional planning approaches failed to meet the needs of the rural poor. It was introduced as a planning approach to help minimize existing investigatory biases, provide an alternative to the limitations of questionnaire surveys, and to give timely information for externally driven planning. PRA built on the principles and methods of RRA but added a new emphasis on enabling local people to undertake their own appraisals, to analyse and act on them, and to monitor and evaluate local changes. Both approaches use similar methods but differ in their purpose and process. RRA is mainly used to collect information and enable outsiders to learn. By comparison, PRA places emphasis on facilitating local processes of learning and analysis, sharing knowledge and building partnerships among individuals and interest groups for local-level planning and actions. Consequently, it is a much longer and open-ended process.

*Source:* Guijt and Hinchcliffe (1998)

---

- Maps, models and transects can be used to locate particular resources. When developed with elders, these can aid an understanding of historical changes in the use and status of resources.
- The values of natural resources can be elicited using a variety of matrix scoring and ranking techniques. These reveal not only how valuable different resources (eg tree species) are to different people but, also, the ways in which they may be important (including non-financial values) and their relative importance compared to other resources and activities.
- Product-flow diagrams and tenure maps can be used to understand how resources and access to them are controlled, and to clarify who is and is not involved in their use, harvesting and management.

The usefulness of participatory approaches is also determined by the attitude of mind and behaviour of the professionals towards the people with whom they work. Success comes from rapport, dialogue and a fair sharing of information and ideas – which means that the professionals too must have attractive trade goods and must appreciate what they are getting in return.

Pretty et al (1995a) give details of how to train in these methods and IIED's PLA Notes (published regularly, three times a year, by its Sustainable Agriculture and Rural Livelihoods Programme) report on how they have been used and adapted in the field.

As participatory approaches have gained sway, attempts have been made to move away from sectoral development projects to more integrated programmes; and also to go beyond the use of participatory techniques just in the appraisal stage of development activities. A good example comes from the coastal region of Kenya where the Kilifi Water and Sanitation Project has followed a participatory and integrated development approach which includes PRA but is, also, concerned with project approval and implementation. The steps of this

**Table 3.4** *Steps in the participatory and integrated development approach (PIDA), Kilifi District, Kenya*

| Phase | Step |
|---|---|
| Initiating phase | 1 Application for PIDA<br>2 Identification of organizing agency |
| PRA phase | 1 First meeting in the community<br>2 Selection of development agencies, forming a PRA team and logistical arrangements<br>3 PRA training<br>4 Second meeting in the community<br>5 Planning the village workshop<br>6 Conducting the village workshop<br>7 Report writing – by the PRA team<br>8 Evaluation of the PRA process |
| Follow-up phase | 1 Draft project proposal (based on the Community Action Plan and the village workshop report)<br>2 Discussion within the community and with other parties involved on the project proposal<br>3 Project application (through the channel: Sub-locational, Locational, Divisional, and District Development Committee<br>4 Parallel to 3: Channelling information for soliciting support and funding<br>5 Parallel to 3: Initiating other development activities, identified through the PRA village workshop within the community<br>6 Operational planning of the approved project<br>7 Project implementation<br>8 Monitoring and evaluation |

*Source:* Schubert et al (1994)

process, which begins with the community requesting its initiation, are listed in Table 3.4.

## PARTICIPATORY PLANNING

Participatory planning is now promoted as an alternative to top-down planning but it faces problems of its own – undefined lines of authority, a weak information base and an institutional culture at both policy level and within organizations that is not conducive to participatory processes. Generally applicable methods of natural resource survey and planning have yet to evolve. They have to win both local legitimacy and recognition by central authorities.

Progress on the ground is most likely to occur where there is an equal and long-standing partnership between land users, planners and natural resources specialists. Box 2.7 in Chapter 2 illustrates a step towards participatory planning in Tanzania taken after more than ten years involvement of the British Land Resource Development Centre in resource assessment and land use planning in Tabora Region. Box 3.7 illustrates how a superficially similar procedure, also in

---

**Box 3.7** *Different perceptions of participation: An example from Tanzania*

Under the Town and Country Planning Ordinance, villages can receive legal title to their land only on completion by government officials of a land use plan, showing what land is to be used for residence, agriculture, grazing, etc. This is then gazetted and becomes legally binding. Following the official planning of Dirma village in Arusha Province, an independent rapid rural appraisal (RRA) assessed the effectiveness of the planning procedure.

The officials had invited leaders from neighbouring villages to discuss and agree the boundaries, which were mapped accordingly. Some conflicts were settled by voting in the Ward Development Committee. Some village leaders had also been consulted to help identify where settlement, farming and grazing should take place, so this was considered to have been participatory planning.

Water is the limiting natural resource. Its availability determines land tenure, settlement, land use and migration yet the land use plan merely stated that each settlement should have water – without specifying ways and means. Unrealistic proposals included zero grazing and the introduction of grade cattle based on 30 acres of pasture per household and a maximum of 18 livestock units (but no proposals for destocking); irrigated agriculture without adequate water supplies; and tree planting (bizarre given the 20,000 acres of *miombo* forest in the village area). Settlements were shown where no one lived, and the area under cultivation was estimated at three times the actual area.

Villagers told the RRA team that, while they had checked the boundaries thoroughly, they left the land use plan to the experts and thought it was merely a formality for securing title. The RRA also revealed that the village leaders were Party appointees and represented mainly the permanently settled villagers, whereas the majority registered in the village are semi-nomadic pastoralists who do not participate in Party affairs, who had not been told of the plan to privatize grazing land, and who were afraid that they would lose their right to move in and out of the village in search of water and grazing.

This kind of prescriptive replacement of all existing agricultural and pastoral practices by modern methods does not build on local people's knowledge of their resources; it could not be implemented without force; and forcing upon the villages would damage both their economy and the environment, and confidence in their own government.

*Source:* Johanssen and Hoben (1992)

---

Tanzania, can fail completely in the absence of interactive participation and adequate natural resources data.

## Examples of local-level resource planning

Many programmes in Africa are promoting community wildlife management. One of the best known can be found in Zimbabwe, where the Communal Areas Management Programme for Indigenous Resources (CAMPFIRE) is devolving power over wildlife and other resources to local people. Box 3.8 gives an example of the approach. Its success has been attributed to the tangible benefits from wildlife that now accrue to the local communities. These benefits have generated increased local support that has enabled the programme to embrace

---

**Box 3.8** *Community wildlife management in Mahenye Ward, Zimbabwe*

Mahenye is a collection of villages extending over about 600 km² on the border of the Gonarezhou National Park in the south east of Zimbabwe. Most of the people were relocated there to make room for the national park. At independence in 1980, rural people were desperately poor, poached as often as they dared, and were extremely hostile to the wildlife, particularly elephants which raided their fields.

To counter this situation, in 1984 the government permitted safari hunting of elephant and buffalo migrating out of the park. The dividends of this hunting were channelled into the local community which was given security of tenure and the right to manage its wildlife resources in the long term. Today, the people have their own committees and other internal government structures and they make responsible decisions. Ten years after the scheme was initiated, there is a welcome resident population of over 300 elephants in Mahenye.

At a meeting of the Mahenye community in 1995, attention was drawn to the fact that a man from the neighbouring community, a successful entrepreneur who owned 500 head of cattle, was in the habit of grazing his cattle on Mahenye land that had not been used for anything else. What were they to do? They could charge him a nominal rent, or ask him to take his cattle elsewhere and put that land under wildlife. They decided unanimously to put it under wildlife.

Mahenye is an example of how decision-making by consensus can lead to the sustainable use of wildlife resources to alleviate rural poverty.

*Source:* Murphree (1995)

---

other communal resources such as grazing, water and woodlands. Despite the acclaim it has received, CAMPFIRE has not been without problems. It has to operate within a confusing institutional framework that has undermined traditional and clearly understood structures (Box 3.9). An independent assessment (PlanAfric, 1997) found that impact on individual household income has fluctuated considerably but has, on average, been low and insufficient to compensate for damage done by wildlife. Inevitably, it noted that there have been examples of financial irregularity by some individual community leaders; and very little infrastructure could be attributed to CAMPFIRE – and some of that was found to be poorly planned (eg silted-up dams). There has undoubtedly been an increase in the capacity to plan and implement projects at the community level, but the rapid expansion of CAMPFIRE has meant that training resources have been thinly spread.

The Aga Khan Rural Support Programme in India and Pakistan has used community-level analysis of problems to determine priorities and investment of both local and external resources. The programme has concentrated on strengthening existing village institutions or establishing new ones as a vehicle for its work. In Pakistan, more than 2000 village organizations have been formed, embracing 75,000 households (World Bank, 1995).

In India, village institutions supported by the programme (established there in 1985) have undertaken diverse activities like water resource development (lift

**Box 3.9** *Institutional framework for development in Zimbabwe*

In 1984, in an effort to increase the involvement of local communities in the planning and development of their communities, the government of Zimbabwe introduced an institutional framework for local development.

Village Development Committees (VIDCOs) were designated as the fundamental planning units and it was envisaged that each VIDCO would represent 100 households (approximately 1000 people). The VIDCO would submit plans annually to the Ward Development Committee (WADCO) representing six villages. The WADCO would coordinate the plans from all VIDCOs in its jurisdiction and submit this ward plan to the District Development Committee (DDC) which would incorporate ward plans into an integrated district plan for approval by the District Council and subsequent submission to the Provincial Development Committee.

By 1989, the Minister of Local Government, Rural and Urban Development stated:

> 'What is disturbing is that in some areas there is an unacceptable level of participation in the planning process by residents at the village and ward levels. Reports reaching my ministry suggest that people are not sufficiently involved or active in the village and ward development committees. *They are not being effectively mobilised to actively participate in development committees in order for them to identify, prepare, and plan their development needs.*' (our emphasis)

There are several reasons for this apathy. First, much of the legislation enacted since Independence has extinguished traditional leadership. The chiefs, sub-chiefs, headmen, and kraal heads are the effective communal lands administrative and legal institutions, with historically defined areas and sets of rules that are clearly understood by the rural people. There is no place for them in the new institutional structure. Predictably, the transition from traditional and chiefly authority (local, hereditary, and long-standing) to elected and bureaucratic (transient and possibly immigrant) has been a source of conflict.

VIDCOs and WADCOs were perceived as instruments of local administration. Ostensibly representative of the rural populace, having been democratically elected, they were essentially implementation units for plans that continued to be developed in a top-down fashion.

Moreover, VIDCO and WADCO boundaries were not necessarily aligned with legal boundaries between communal lands, thereby creating uncertainties over institutional jurisdiction (a situation further complicated with the amalgamation of Rural and District Councils in 1993/94 when some wards were merged while others were sub-divided). Administrative structures within many communal lands remained confused, at least from the perspective of the inhabitants.

The government's intention to train VIDCOs and WADCOs in administrative skills has proved over-ambitious, not least because of a lack of sufficient financial and human resources. VIDCOs and WADCOs have tended to operate, if at all, in a vacuum, without the wherewithal to enable them to function effectively and with no mandate from their constituency. Rural people have had little option other than to get on with their lives, much as they did before independence. The air of optimism at the start of the 1980s has given way to resignation.

The above criticisms of VIDCOs and WADCOs appeared in the reports of the Rukumi Commission on land tenure (Government of Zimbabwe, 1994) and were partly accepted by parliament – particularly in terms of the desirability of re-investing the traditional leaders with some of their former powers. A Traditional Leaders Act (TLA) was passed in 1998 which sought to make the old WADCOs and VIDCOs elected committees of new structures – Ward and Village Assemblies (led by traditional leaders). The functions of the VIDCO remained as described in the Rural District Councils Act and those of the WADCO, previously undefined, were set out in the new LTA. However, the provisions of the LTA remain to be implemented by November 2000 due to a lack of resources.

*Sources:* Thomas (1995); Derek Gunby, PlanAfric consultants (pers. comm.)

pumps, community wells, percolation tanks, check dams, minor canal irrigation); reforestation of degraded forests; agricultural extension; savings and credit; and group marketing. Reforestation and protection of degraded common lands has been identified as a priority by a number of villages in the Bharuch District. A total of 2500 ha of land had been reforested and protected by the village institutions in this district by 1993 (Shah, 1995). The *Participatory Rural Appraisal and Planning* approach is supported by the professional staff with the interactive participation of the local people. The process starts in a village with the preparation of a natural resource inventory by the people using maps, transects, time lines and seasonality analysis. The resources are analysed in terms of their use, productivity, ownership, status and the access people have to them. Social analysis is carried out at the same time using social mapping and wealth/well-being ranking. Social and economic information is then analysed along with the resource inventory and classification of resource users/owners to identify different focus groups within the community. Focus groups separately analyse the status of their resources and the problems/constraints they face. Each group prepares a resource-problem-opportunity matrix, listing different problems in using each resource and the solutions to these problems, and arrives at its priorities for action.

The various analyses are then discussed in a village meeting and a village plan is negotiated by prioritizing the different concerns presented by the groups. Figure 3.3 illustrates the process in Bharuch District.

## Scaling-up and linking bottom-up and top-down planning

### Regional rural development

There have not been many examples of successful scaling-up of local planning initiatives to the national level. One response from the German development agency GTZ in the 1980s, is known as Regional Rural Development (RRD):

> '*The concept entails giving the broad bulk of poorer population strata the capacity to make better use of the resources available to them with the aid of support measures adapted to specific situations and established on a participatory basis, and consequently providing them with an opportunity to become involved in the process of regional development*' (GTZ, 1993).

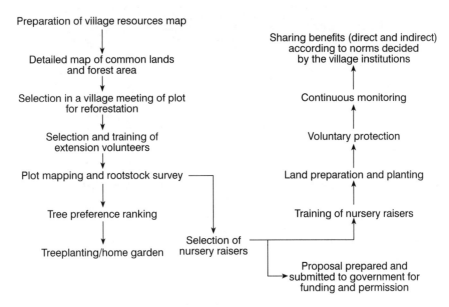

*Source:* Shah (1995)

**Figure 3.3** *Participatory rural appraisal and planning (PRAP) process Bharuch District, Gujarat, India*

RRD encompasses two principles: a focus on poverty with special emphasis on the participation of the poor, and sustainability, combined in the overall goal of sustained improvement (or stabilization) of living conditions. All other development policy targets, principles and guidelines included in the RRD concept are linked to this goal in a hierarchical system of objectives shown in Figure 3.4.

RRD particularly aims to ensure the participation of disadvantaged *target groups* in the planning and decision-making process. It promotes procedures which are practicable under local conditions and within the capacities of the regional or district planning offices or agricultural advisory services.

There is an emphasis on *participatory dialogue* so that decisions are mutually coordinated between development programme sponsors and the beneficiaries. The procedures are based on *free choice* which implies a restriction on government decision-making authority concerning the use of public funds and channelling of private-sector economic activity. GTZ (1993) suggest that target groups could participate through a range of instruments:

- *Un-earmarked grants* to village communities for communal activities – a far-reaching means to strengthen self-determination and, in practice, decentralization to the very local level.
- *Multi-purpose funds* from which government complementary services for local self-help could be provided on application – allowing communities to select from a predetermined range of supported projects.

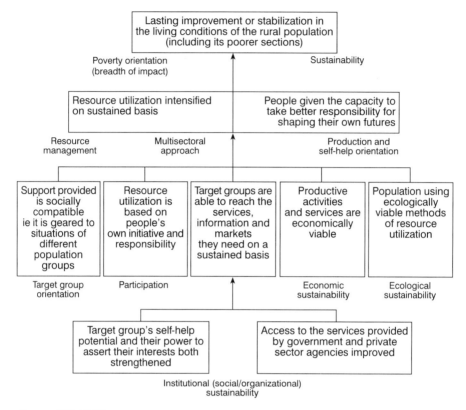

*Source:* GTZ (1993)

**Figure 3.4** *The regional rural development (RRD) system of objectives*

- *Un-earmarked loans* to individuals or groups – allowing them to make an autonomous decision on how funds are to be used.
- Governments could offer a *broad selection of inputs, means of production and recommendations* that return some of the responsibility for production planning to small producers themselves (allowing them to organize matters in their own way and to experiment).
- Measures to *improve local transport facilities* provide farmers with greater opportunities of access to larger input and output markets and broaden the available choices.
- Measures to *improve rural information services*.

Even mechanisms which only increase people's decision-making indirectly need to be included in a strategy to increase target group participation. As GTZ note:

> *'If (s)he can personally travel into town to buy seed, this gives the small farmer a lot more decision-making power than if a non-elected representative joins in the discussions on what type of seed the marketing authority ought to provide to the villages, without even having consulted the people (s)he is supposed to represent'.*

Target group participation is expected to increase the chances that:

- problem analysis during development planning will reflect the way problems are viewed by various groups (particularly the disadvantaged);
- the decision-making processes will take account of solutions put forward by the target groups and their assessment of the alternatives under discussion;
- target group activities and governmental support measures are mutually coordinated;
- the government monitoring system systematically records how the measures are judged by the target groups;
- target groups are aware of, and are using, the ways and means available to them for making their needs, complaints and suggestions known to the responsible bodies.

No one standard procedure for participation is laid down. Table 3.5 indicates how these approaches might be deployed in different stages of the planning process. GTZ (1993) lists various options for engaging target groups which can be used, as appropriate:

- questionnaires;
- institutionalized (and formalized) information channels between villages and those responsible for programmes/projects (eg through representatives of target group organizations, the extension services or political representatives);
- informal group discussions or one-to-one conversations;
- formal gatherings (eg meetings of self-help groups, cooperatives or associations; official village meetings);
- official application procedures, eg for support to local infrastructure;
- joint agreements;
- participatory action research, in which target group representatives are involved in establishing what compatible measures should be carried out locally (one of the participatory methods in the PLA family – see 'Participatory learning and action', page 98);
- participation in government planning events by target group representatives.

GTZ argues that in order to achieve a dialogue with target groups, planners must go into the villages, and individual planning stages must be relocated into the villages.

## Rapid district appraisal (RDA)

To date, participatory approaches have been used quite successfully at the local level. Practitioners have also been concerned about the question of 'scaling-up' participatory approaches for application at broader and higher levels, particularly at the district level. The district environmental action plan (DEAP) process in Zimbabwe (Box 2.12) is strongly based on local participation and ownership. In Indonesia, the Institut Teknologi Bandung has introduced a postgraduate training programme for regional planning which aims to promote rapid district appraisal (RDA) as a broader-scale application of RRA (Box 3.10).

**Table 3.5** *Forms of participation for RRD at different stages of planning*

| Forms of participation | Questionnaires | Voicing of positions without having to ask | Institutionalized information channels | Conversations and group discussions | Meetings empowered to make proposals and applications (target group level) | Participatory action research | Participation in planning meetings at government level | Official application procedures | Written agreements |
|---|---|---|---|---|---|---|---|---|---|
| **Stages in planning** | | | | | | | | | |
| Situation analysis | 1 | 0 | 0 | 1 | 0 | 0 | 0 | 0 | 0 |
| Objectives analysis | 2 | 0 | 0 | 1 | 0 | 0 | 0 | 0 | 0 |
| Problem analysis | 1 | 2 | 2 | 1 | 1 | 2 | 2 | 0 | 0 |
| Consideration of alternatives | 0 | 0 | 0 | 2 | 1 | 0 | 1 | 1* | 0 |
| Planning of measures | 0 | 0 | 0 | 1 | 2 | 1 | 1 | 1 | 0 |
| Operations planning | [–] | [–] | [–] | 2 | 0 | 0 | 2 | 0 | 1* |
| Monitoring, feedback, plan adjustment | 1 | 1 | 1 | 1 | 2 | 1 | 0 | 0 | 0 |

*Notes:* 1 = very important, [–] = not worthwhile, or impossible, 2 = supplementary importance, * = for self-help promotion projects, 0 = unlikely in rural development
*Source:* GTZ (1993)

---

**Box 3.10** *Rapid district appraisal (RDA) in Indonesia*

RDA seeks to understand district conditions using a mix of methods: written, verbal, visual and interactive. The application of RDA can be wide ranging, from medium- or long-term spatial planning (in the Indonesian context, the five-year planning framework – REPELITA), to annual planning (as in the Indonesian REPETADA), to policy research, environmental assessments or programme evaluations.

Participatory appraisal of all settlements of the district is, clearly, an impossible task. Instead, RDA structures a district according to natural and/or socio-economic features, and selects villages representative of each zone. *Case studies* are then undertaken in these villages. The approach follows the concept of *recommendation domains* – areas defined as homogenous according to, for example, agro-ecosystems/ resource endowments, topography, accessibility and socio-cultural grouping.

The initial steps are a collection of secondary data, field reconnaissance and preliminary analysis to yield a general picture of the conditions in the district, based on which recommendation domains can be established. If the necessary data are available, GIS could be used. Alternatively, a representational model could be built and the main natural and socio-cultural features discussed and verified with groups of local experts. To help select representative sites, it is necessary to define in advance the issues to be analysed and indicators for these.

Sub-districts or villages can be selected which highlight key issues and problems which lend themselves to particular RDA methods. For example, particular sub-districts may have a high incidence of poverty. Further examination of secondary data might reveal that there are at least two agro-ecosystems in the area in question and that some of the villages are located in hills and isolated from roads. In this situation, two contrasting villages could be selected: a typical isolated hill village and one from a different eco-system without such locational problems. The causes of poverty in each will probably be somewhat different and require different solutions.

RDA uses a basket of methods, but not in any standard sequence, including:

- mapping, preferably with or by key informants from the area;
- developing a 3-dimensional model of the region;
- transect walk or transect sketch;
- semi-structured interviews;
- focus group discussion(s);
- institutional diagramming;
- local or regional feedback meeting.

*Source:* Kievelitz (1995)

---

## Participatory approaches in large-scale projects

There is little significant and documented experience of the transfer of participatory approaches to hierarchical organisations and large-scale projects. However, one example is provided by Sri Lanka's North Western Province Dry Zone Participatory Development Project (Box 3.11) which illustrates some of the problems that arise when transferring the participatory approach from one institutional culture to another.

---

**Box 3.11** *The North Western Province Dry Zone Participatory
Development Project (DZP), Sri Lanka*

The DZP is a large investment programme funded by the International Fund for
Agricultural Development (IFAD) and GTZ, implemented through provincial government
agencies and coordinated by the Regional Development Division of the Ministry of
Finance and Planning. It aims to facilitate participatory planning in 500 villages (located
in 13 administrative divisions) over a seven-year period and establish Village Resource
Management Plans for each of these villages. The government services can use these
plans to assist poor farmers by providing technical advice and funding for the resource
management activities selected by the farmers. The project assistance, however, is
limited to a list of predefined project components such as the (small-scale) development
of water resources for irrigation, upland farming systems development, goat rearing,
land regularization and credit.

Eighteen months after the inception of the project, participatory activities had taken
place in about 40 villages.

High expectations were raised during the participatory fieldwork but government
service officers have found it difficult to fulfil promises. Also, after PRA training, people
tend to revert to their old hierarchical social system. Tacking on PRA to field officers'
methods doesn't necessarily lead to fundamental changes of attitude or better rapport
with beneficiaries – indeed some officers were tempted to invent PRA results.

> 'Government agencies cannot be expected to implement a participatory
> project successfully and instantly. An orientation or transition phase (which
> might require two to three years) is needed to enable staff to learn and to
> adjust, and for strategies to be developed and tested. The adoption of a
> participatory working style in a hierarchical organisation has to be a
> continuous step-by-step process. It requires experienced and qualified
> people to facilitate the process of discovery and learning. Formal staff
> training, although important, is not sufficient. Continuous backstopping
> and coaching are more suitable. For this process, the usual short-term
> inputs of consultants and trainers are of limited usefulness. What is
> required are persistent "changes agents" coming from outside the
> organisation who are available over a longer period of time.'

*Source:* Backhaus and Wagachchi (1995)

---

## The catchment approach

Success in adopting participatory approaches requires a long-term commitment.
In Kenya, for over a decade now, the Ministry of Agriculture has pursued an
interdisciplinary catchment approach to soil and water conservation, seeking to
involve all interested parties at local level (both resource users and external
government and non-government agents) in planning, decision-making,
implementation and maintenance (Box 3.12). The sheer time, effort and cost
required to establish and maintain the community structures is only realized
when the effort is made – as revealed by a similar approach for water catchment
planning now being followed in Zimbabwe. A National Water Authority has
been created, and catchment and sub-catchment Councils established
throughout the country with stakeholder representation, including local

---

**Box 3.12** *The catchment approach to soil and water conservation, Kenya*

In Kenya, soil conservation is the responsibility of the Soil and Water Conservation Branch (SWCB) of the Ministry of Agriculture, Livestock Development and Marketing which operates in 222 divisions in all 47 districts of the country. The *catchment approach* to soil and water conservation was adopted in 1988 upon realizing that conventional approaches (through farmers being advised, lectured, paid and forced to adopt new measures and practices) were not effective. The objective was to concentrate resources and efforts while ensuring participation of the community within a specified area (typically 200–500 ha) for a limited period of time. In this approach, the term catchment is closely associated with a specific community of people known to each other, rather than a physical unit.

Local communities are purposefully involved in the analysis of their own farming and conservation problems, and decisions and recommendations made with their active participation. Community mobilization is achieved by interdisciplinary planning teams, the formation of catchment conservation committees by farmers themselves, and intensified publicity and training through field days, public meetings, demonstrations and tours. This process enables information to flow to the community, the development of better understanding of the conservation problems specific to each area by the SWCB, and closer collaboration between farmers, the SWCB and other agencies.

Each divisional planning team (DTP) typically works in three or four catchments each year. Priority is given to catchments where local people or administrations have requested support, where soil erosion is serious, or where the SWCB has not worked before.

Multi-disciplinary teams drawn from various government departments work for about a week in the catchment: beginning with a day of orientation and introduction to the methods, followed by 2–3 days building up a picture of local skills, knowledge and perspectives on problems and concerns using a variety of participatory inquiry methods, with a public meeting on the final day to present the findings in visual form.

Following the dialogue, a catchment conservation committee (CCC) of farmers is elected to coordinate action within the catchment (typically 8–15 people with the local Technical Assistant as an ex-officio member). The divisional team then prepares a catchment report to serve as a baseline document for planning, implementation, monitoring and evaluation, and for the coordinated action by extension professionals based at divisional and district level. The DTP makes a detailed map of the catchment, and plans the soil and water conservation measures for each farm, working with the catchment committee. The CCCs receive support in the form of basic tools, equipment and technical training and advice from ministry staff. In return, committee members assist fellow farmers in planning and implementing various individual and group soil and water conservation activities, and support the DTPs in laying out and implementing plans for each farm.

The success of the approach is, in part, due to the ownership of the plan and commitment that has been achieved by active local communities (through CCCs) working with the professional teams. Lines of communication have been established and farming communities can exert a pull on the services of extension agents.

Kiara et al (1999) report growing evidence of positive impacts arising because of the catchment approach. They compare, for example, two catchments in Trans Nzoia, one planned with the catchment approach and PRA and the other through the contact farmer individual approach. Crop yields and returns per person per day have grown more rapidly on the farms in the community where the catchment approach was used. As a result, land values have increased dramatically, together with some increase in leasehold prices. Farming has become more diverse in the catchment approach area with a greater range of crops grown and more livestock kept.

By the mid-1990s, some 100,000 farms a year had benefited from the catchment approach; Each year, some 500,000–800,000 m of cut-off drains and 50,000–100,000 m of artificial waterways are constructed, some 1250–2700 gullies controlled and 1780–3600 km of riverbanks protected.

The SWCB was proposing in the 1995–96 financial year to launch participatory planning in 809 catchments covering 177,000 ha and 93,000 farm families (John Thomson, quoted in Chambers, 1995).

Some impact studies carried out by MOALDM-SWCB show that in the conventionally planned catchments, the process begins with a *baraza* (community meeting), which is held for publicity purposes. The catchment committees are sometimes elected but, more frequently, are selected by chiefs or local leaders. Women are rarely represented, and farmers are not involved directly in planning and layout. The committees tend to become inactive soon after intensive contact with extension staff ends.

However, where the catchment approach mobilizes the community, supports strong local groups and brings in committed local staff and collaboration with other departments in interdisciplinary planning and implementation, there is increased agricultural productivity, diversification into new enterprises, reduction in resource degradation, enhancement of water resources, improvement in the activities of local groups and independent replication to neighbouring communities within two years. The land treatment plans developed in consultation between farmers, catchment committees and DPTs embrace broad-based recommendations, especially for soil fertility: unlike the former emphasis on physical erosion control. These improvements have occurred without payment or subsidy and, therefore, are more likely to be sustained.

A recent review of the catchment approach and its contribution to capacity development for soil and water conservation provides a summary of the approach (Harding et al, 1996): the major actors, HQ staff, the district and other staff, and the farmers and farming communities, were linked by what might be called virtuous cycles that reinforced the exchange and linkage between policy development, practice in the field and training at all levels. This was at the heart of the learning culture. The factors which helped to drive the cycles (which to some extent also turned themselves) included: a strong political push from the government; Sida as flexible funders and facilitators; the Regional Soil Conservation Unit providing critical assessment and broader views; and additional external support from organizations such as IIED, providing training and advice. This review also notes difficulties in implementing the approach. But these are well recognized by SWCB which is constantly re-assessing and re-working it to deal with them. The report concludes, among other things, that the Kenyan experience 'illustrates well a hard reality of much development work; that all progress is partial and qualified, and that the impact of constant change forces continual re-assessment of even successful approaches'.

*Sources:* Pretty et al (1995b); Harding et al (1996)

communities. The example of Mazowe catchment – one of the pilot catchment planning initiatives – is described in Box 3.13. In Tanzania, the HIMA programme in Iringa Region also adopted a catchment approach from the outset.

---

**Box 3.13** *The Mazowe Water Catchment Planning Pilot Project, Zimbabwe*

### Establishing the structure

Land and water users in Mazowe catchment, Mashonaland, had a vision for some form of user board which embraced all water users and covered the whole catchment. Following a workshop held in July 1996, a working group was formed to set up Water User Boards (between ward and district level), each nominating two representatives to Sub-Catchment Councils (between district and provincial level) each, in turn, nominating two representatives to a Catchment Council. The process took a year to build awareness, trust, participation and institutions that are transparent and have representation for a range of user sectors, eg communal areas, mining, industry and agriculture. Civil servants and other experts act as advisers.

### Building the lower tiers

The Mazowe Catchment Council has undertaken a lot of work including establishing the lower-tier structures (Sub-Catchment Councils and Water User Boards). Educating the tiers of structures about their responsibilities is a challenge and attendance at meetings is erratic, especially for members living in remote areas, due to poor communications and the cost of travelling.

### Publicity and awareness campaigns

The Catchment Council has embarked on an awareness campaign to inform water users about water sector reforms and activities in the catchment area. Publicity materials, prepared by the Water Resources Management Strategy (WRMS) project in English, Ndebele and Shona, have been distributed. But there is a constant demand for more and efficient distribution of these materials – another challenge.

### Catchment plan

It is taking a long time to prepare an integrated plan for the catchment. The task has been found to be more complicated than originally anticipated.

*Source:* PlanAfric (2000)

---

## NGOs as catalysts

NGOs have been important catalysts in the development and application of participatory planning at local level and, also, at national level. For example, in 1992, the Mexican government delegated preparation of the natural Tropical Forest Action Plan to its strongest critics led by a consortium of NGOs. There are other cases where NGOs have assisted local communities to identify their own priorities, and State representatives participating in these discussions have suggested ways and means to carry out the plans. Pooling of community priorities at regional or national level enabled them to link with the development of policy. In Papua New Guinea, an almost unprecedented opportunity has been created for NGOs and customary landowner groups to participate in decisions about the management of national forests as part of the development of the National Forestry and Conservation Action Programme (see Mayers and Peutalo, 1995). And in India, participatory methods have now

> '*spread well beyond the confines of the NGO and academic circles where they were developed and where their use was characterised by innovation and flexibility. PRA methods have become part of guidelines for major state initiatives, such as the new national watershed development programme, in which speed, scale and bureaucratic management give shape to their use*' (Mosse, 1995).

## *The* gestion de terroir *approach in francophone West Africa*

Following the 1984 Regional Conference of Sahelian States in Nouakchott, most countries in the region initiated national plans to combat desertification, several of which adopted an approach known as *gestion de terroir* (GT). Two terms, *gestion de terroir* and *aménagement de terroir*, are commonly used, often synonymously, to describe the range of community-focused projects. Terroir refers to a socially defined space containing resources and associated rights, within which a particular community is assumed to satisfy most of their needs. *Gestion de terroir* refers to the management of natural resources: allocating land to certain uses, limiting access at certain times, and controlling levels of resource use. *Aménagement de terroir* refers to the improvement of resources, involving a variety of investments to raise productivity, reduce crop risk, and conserve soil and water.

Over the course of the 1990s, the GT approach to local land use planning has been adopted by myriad government projects, donors and NGOs operating in the Sahel. In 1994, in Mali alone, over 200 GT projects were engaged in natural resource management, setting up local land use committees to establish land use plans, and financing complementary development activities to encourage the adherence of the local people (UNSO, 1994).

The driving principle of GT is the devolution of decision-making powers for land use planning and natural resource management from government agencies to the local people. This stems from the recognition that governments are not well equipped to manage land at local level, and that local people often have both sound technical knowledge and a range of institutional structures capable of managing resources. There are many variants of GT which can be differentiated by the relative attention paid to either the degree of local participation, or the extent to which they address natural resource management or socio-economic issues (Winckler et al, 1995; Yacouba et al, 1995).

Three broad approaches have been taken by GT projects:

*The natural resources management (NRM) approach* focuses on the physical improvement of the land with emphasis on soil and water conservation and agro-forestry through existing institutions (modern and traditional) as well as with individuals. Initially, the focus was on collective erosion control structures, nurseries, etc. Later, the emphasis shifted to activities managed by individual farmers themselves, such as raising soil fertility. Initial training is provided in soil conservation but, over time, this is extended to broader issues of land use planning and management. The basis of this approach is that local people can see immediate benefits in the form of improved crop yields which should encourage the adoption of conservation techniques by farmers in neighbouring communities.

The main shortcoming is that the projects do not address any other (and maybe more pressing) socio-economic needs of the population.

*The institution-building approach* focuses first on establishing and training community-based institutions which are supposed to design and implement a land use management and development plan. Funding and credit is made available to assist with the implementation of selected activities that meet local priorities (for instance, wells and dams). Such an approach generally consists of two phases: phase 1 involves establishing and training a village GT committee, carrying out a participatory study of village resources and institutions and drawing up a land use management plan. Phase 2 involves implementing this plan through NRM and socio-economic activities and the enforcement of new NRM rules and regulations.

In principle, this approach should establish a village-based institution trained in holistic land use planning and management that is capable of promoting a sustainable development strategy. In practice, however, the capacity of the committee to carry out its mandate is often compromised by a lack of legitimacy: either because it has displaced existing customary institutions, or because it is neither democratic nor representative. Each GT-programme tends to create its own village committee, and several years filled with meetings and making plans can pass before local people see any tangible benefits from their involvement in the project.

*The local development approach* is a more recent adaptation of the two former GT approaches and tries to address their limitations. It also reflects the growing donor and State interest in decentralization and privatization. Three aspects differentiate the local development approach from the institution-building and NRM approaches:

1   Community organization and fund-raising is located at a supra-village level, even though specific activities may be carried out at village or group level. Emphasis is on collective investments.
2   Financial responsibility for rural land use planning and NRM is transferred from the externally managed project to community-based land use and planning committees. These are given control of a credit fund provided by the project to implement its activities and must, also, raise complementary funding. The committee has to set its own priorities for allocating the limited funds available to activities proposed by the various communities.
3   This committee is expected to call for tenders, select and sign a contract with a local contractor to build infrastructure and implement other development activities. It must also monitor the performance of contractors.

If the local community is given power to decide how to allocate funds, there is the risk that community organizations will give priority access to relatively capital-intensive investments, so as to be able to absorb the available funds. They may also succumb to local pressures to invest in social infrastructure activities to address immediate priorities (such as a village water supply) at the expense of investing in longer-term, sustainable NRM activities. Some projects have reacted by earmarking funds (eg 70 per cent on NRM and 30 per cent for other issues) or requiring that these investments are linked to NRM (eg treeplanting around a well). Issues related to management and regulation of natural resources are likely to receive less attention in the local development approach, particularly when accompanied by large funds that have to be invested locally.

In theory, these approaches promote an integrated and participative process of rural land use planning and natural resource management founded on the involvement of local people according to customary practices. In reality, however, they have not fully lived up to expectations (Box 3.14).

## Participatory planning in Latin America

The World Resources Institute has studied several cases in Latin America where, first, local priorities were identified by local communities with assistance from NGOs. State representatives participated in the discussion of local plans, reacted to ideas emerging from communities and provided information and ideas about means of carrying out the plans and about potential constraints. Pooling of community priorities at regional or national level enabled them to link with the development of policy (Lori Anne Thrupp, pers. comm.). Of course, such linkage depends on a willingness to empower local institutions and the building of a planning framework in which it can occur. In Mexico, Colombia and Chile, laws have been passed requiring the establishment at the provincial or local level of environmental planning committees with broad social representation (Zazueta, 1995) and governments are increasingly delegating the planning and implementation of programmes to independent sectoral organizations, eg the Mexican Program for the Protection of the Tropical Forests (PROAFT) (Box 3.15).

## Approaches in the forestry sector

The forestry sector has undergone a sea change over the last 10–15 years, from production forestry to management forestry. In many countries, there have been many initiatives in collaborative resource planning and management which include social forestry, community forestry, joint forestry management, participatory natural resource management, environmental stewardship, co-management of protected areas and integrated conservation-development projects. Nepal's community forestry programme is well-known and frequently cited as a good example of successful participatory forest planning and management. Originally concentrating on involving local people in the management of new plantations on degraded land, the focus has now shifted to

**Box 3.14** *Some problems with the* gestion de terroir *approach*

## Policy vacuum

Despite a general commitment to the GT approach, most Sahelian governments did not articulate a specific policy for the manner of its implementation (UNSO, 1994). Burkina Faso is a notable exception: the government made GT a specific policy in their efforts to promote rural development, and created a national land use planning and management programme to ensure a harmonized and coordinated approach (*Programme National de Gestion de Terroir*) within the Ministry of Agriculture. Elsewhere, government departments set up ostensibly to coordinate and monitor the implementation of GT projects were largely ineffective. This situation may be explained for instance in Mali, partly, by insufficient funding, lack of political will, poor communications across government departments and independent-minded donors and NGOs promoting GT projects. Resistance was encountered by other organizations that took the initiative to stimulate exchange and coordination among the multitude of NGOs, donor-led projects and parastatals involved in GT programmes (CMDT, 1991). As a consequence, GT programmes operated in an institutional vacuum.

## Technical focus with a pre-established methodology

Most GT projects have tended to focus on the physical aspects of NRM, ignoring the more complex social, economic, political and cultural factors that determine how households can most effectively use these resources. GT projects are frequently seen as a technical exercise to produce a plan for several years, by implementing a pre-established series of steps (extensively described in project manuals) to be taken by project field staff. Emphasis is often put on the preparation of maps (soils, vegetation, land use zones) which generally are produced by project staff. In contrast the activity plan sometimes resembles a 'wish-list' to be presented to potential donors.

GT is seldom regarded as a planning *process* undertaken by local stakeholders to assess and negotiate resource use, establish rules, regulations and land use practices to better manage their resource. There is seldom an emphasis on developing contingency plans to allow for changing circumstances such as drought, government policies and world markets.

In the analysis of land use problems, GT has, also, tended to focus on farming – at both household and village territory levels. Individual farm holdings are seen as autonomous, technical entities and this is not conducive to understanding the complexity and diversity of rural production strategies. The preoccupation with technical issues and the assumption of uniform community interests has neglected the socio-economic and cultural heterogeneity of communities. Differences in access to land, labour and credit, resulting from village power relations, gender, age and caste tend to be overlooked (Painter, 1993; Painter et al, 1994). Committees are dominated by local elites to the exclusion of certain groups (see below).

## Misapplication of the methodology

The formal procedure for the implementation of GT is to entrust the local community with the authority to carry out its own analysis of the situation and design a plan to meet its priorities. In practice, however, many GT projects adopt a top-down approach.

Degnbol (1996) argues that most GT projects have been implemented as conventional extension packages in a political and institutional climate which does not seriously question or redefine the relationship between the State and the local population. In addition, project staff who are in direct contact with the population, and who are responsible for implementation of the GT activities, have a limited basic education and receive little training in participative methods. They may also face transport constraints, while others link their involvement to the *per diems* they will receive. Mobility and flexibility of government field staff can also be limited when planning is rather inflexible and personal initiative is not valued: the use of participatory methods has seldom resulted in changes in the way institutions behave.

### Sectoral approach

GT demands a holistic approach with close collaboration between the various government services dealing with livestock, agriculture, forestry, etc. The local development approach will also explicitly include socio-economic issues such as health care, sanitation and water, education and roads. However, this approach goes against the grain of existing government rural development structures which are organized along sectoral lines with their own distinct areas of responsibility and limited horizontal communication or collaboration. A promising development in Mali is the restructuring of the Ministry of Agriculture and Water which started in 1997. The livestock, forestry and agriculture divisions have been dissolved and replaced by three new divisions focusing on 'community organization', 'rural equipment and land development' and 'legal issues and control'.

### Lack of a rural finance infrastructure

The absence of agricultural credit banks or community-based credit and savings banks has been a serious handicap. Funding of rural planning and NRM activities has been, almost exclusively, covered by donor and NGO projects. This restricts the role of the community to that of recipient and executor rather than decision-maker and manager. Credit, at the moment, is only readily available in cash-crop-growing areas. A promising development in the 1990s has been the installation of several community-based savings and credit cooperatives like the *kafo jiginew* in the cotton-growing belt of southern Mali.

### Local bias with limited attention for institutional issues or influencing policy

Initially, there was a tendency only to address problems prevailing at the community level. Institutional issues such as land tenure and local rules of access and management of resources were initially rarely addressed. Many projects operated autonomously, even though formally attached to government structures. Degnbol (1996) makes the point that while many GT projects have had some successes in addressing NRM issues at the local level and in improving the livelihoods of those involved, they have had a limited impact in tackling the structural causes of rural poverty or influencing national policies in this domain. In Degnbol's view, rural planning and NRM policy in Mali has so far been mostly influenced by large-donor and government-driven projects, each operating in a specific region. These projects tend to focus on technical and administrative solutions for problems which are fundamentally an issue of power relations between the state and civil society. This lack of interest in the legal, policy and administrative environment began to change in the 1990s when some GT programmes started to address land tenure issues, and supported the decentralization programme.

---

**Village land focus**

The basis of GT is the village land (*terroir villageois*) that is traditionally managed by a village which has recognized rights of occupation and use. This *terroir* focus is often biased in favour of settled communities. It does not easily embrace the interests of nomadic groups. Local land use plans are rarely developed in consultation with transhumant herders who may rely on these resources. The poor integration of seasonal visitors' needs is increasingly recognized by GT programmes, although they still face serious practical difficulties in involving non-resident populations in the planning and implementation of GT activities (Diarra, 1998).

The GT approach has also been criticized for its implicit supposition that the village territory represents the sole livelihood resource for the community. Painter et al (1994) and others argue that Sahelian farmers and agro-pastoralists exploit a far wider 'action-space' than the immediate vicinity of their village. Urban–rural linkages are, likewise, not taken into account.

---

management of natural forest, and user groups can incorporate than own management practices where these are effective (Box 3.16).

## Landcare in Australia

Probably the largest community-based movement is Landcare in Australia, initiated in 1989 by the unlikely alliance of the Australian National Farmers' Federation and the Australian Conservation Foundation. Landcare groups are voluntary associations of rural people who work together and in collaboration with state agencies to look after their own neighbourhoods. Their varied activities, which are almost identical with those of farmer groups in the long-established *soil conservation districts* in the US, include:

- development of a catchment or district plan, identifying major problem areas, and proposals for dealing with them (Box 3.17);
- active involvement in natural resource monitoring, often in conjunction with schools, state agencies and other professionals;
- documenting local knowledge about land and its management;
- study tours of their own and other regions;
- joint research with universities, research bodies, and state agencies;
- production of educational materials.

Landcare groups are concerned with a wide variety of issues: erosion and degradation; water/river-related issues, weeds, nature conservation and biodiversity, education, wetlands, waste minimization, extractive industries rehabilitation, tourist impact management, salinity, feral animals, and both conservation and sustainable farming.

Landcare pays particular attention to *land literacy*. This involves activities which assist people to *read the land*, eg overflights for farmers to see their land and the extent of degradation, publications and kits to assist land users in recognizing emerging problems such as soil and stream salinity.

**Box 3.15** *Delegating planning to NGOs: the case of PROAFT in Mexico*

Like most of its contemporaries, the Mexican Tropical Forest Action Plan (TFAP) had been drawn up with little consultation with local people and was heavily biased towards the forestry sector. Officials within the Ministry of Agriculture and Water Resources recognized that it did not address key policy issues, such as intersectoral policy linkages, the needs of forest dwellers, and the impoverishment of marginal populations, and so could not arrest rapid deforestation and loss of biodiversity in the Mexican tropics.

To address these concerns within the Ministry, in mid-1992, the Undersecretary of Forestry invited some of the strongest critics of the government's forestry policy to propose an alternative. As a result, the top-down Mexican TFAP turned into PROAFT, a highly participatory process that involves stakeholders in planning with action. At a time when the side agreements for the North America Free Trade Agreement were under negotiation, tropical forest conservation became an important concern for former President Salinas. Indeed, he personally conferred decision-making authority to PROAFT.

Over the next three years, a team of five people from two established NGOs (Gestión de Ecosistemas AC, and Grupo de Estudios Ambientales AC) and the National University was assembled and a new NGO (PROAFT AC) was formed to act as a counterpart and procure funds for the three-pronged initiative:

1   A series of 16 tropical forestry studies. Some were generic, such as those on the expansion of cattle herding and legislation on resource use in the tropics, while others were specific, such as performance evaluations of specific commissions or programmes. These studies provided an overview of the condition of natural resources, summarized the outcomes of government programmes in tropical areas, and contained recommendations for improving those programmes or initiating new activities or policies.
2   A series of Tripartite Alliances, through which PROAFT would promote grassroots initiatives to improve forest management by financing and providing technical assistance to projects identified by the communities. NGOs and universities were invited to provide technical assistance to these community groups – thus the name Tripartite Alliances (the Ministry, the community and an NGO or university).
3   A process of consultation through workshops carried out in various regions of the country. PROAFT presented priorities that were debated and amended in open discussions by representatives of NGOs, grassroots groups, business and government officials from other ministries and state governments. On average, 60 people attended each workshop.

Through this process, PROAFT had, by late 1994, produced a new Mexican TFAP that harnessed a broad spectrum of knowledge and viewpoints; identified six priority lines of action, each of which included a series of proposed activities; and identified organizations that could implement them.

Zazueta (1995) reported: 'Mexico's political and economic crisis in late 1994 and early 1995 has placed many government initiatives on hold, PROAFT amongst them. Nonetheless, PROAFT's participatory approach has led the new administration to review and approve the proposal. Because PROAFT is not only technically sound but, also, incorporates the views of the various stakeholders, it appealed greatly to new government officials seeking to respond to the democratisation of Mexican institutions.'

**Box 3.16** *Nepal community forestry programme*

In Nepal, forests were nationalized in 1957, placing them under the control of the Forest Department. It soon became clear that the department lacked the capacity to manage the forests effectively. Regulations made life difficult for the people whose farming system depends on a variety of forest products, browse and compost but people continued to use the forest illegally – they had little choice although, if caught, they were subjected to significant penalties. Control was inconsistent and, to a large extent, the authorities ignored the forests in the hills, except for attempts to police forest use.

In the late 1970s, innovative thinking by a number of Nepali foresters led to a new approach involving handing over forests to local *panchayats* (official politico-administrative units) that were willing to protect them. Legislation allowed for the use and harvesting of forest products by the people of the panchayat, subject to the Forest Department's approval of a management plan. In practice, however, very little forest was handed over prior to the late 1980s, and very little of that was governed by management plans that allowed any significant forest use. In some areas, the establishment of plantations and the protection of natural forest were successful but, with few exceptions, the benefits to people were few, particularly in terms of access to forest products.

The completion of a national forestry master plan in 1988 and the issue of operational guidelines to assist implementation of the plan pending revised legislation (The Forestry Act, January 1993) resulted in easier implementation, increased incentives for people's participation and, consequently, a rapid expansion of the programme. Significant features of the programme as it now stands are:

- Forest management agreements (operational plans) are negotiated between the Forest Department and groups of people with a direct interest in the use of a particular forest and claiming usufruct, rather than with larger administrative units.

- Under the legislation and guidelines, the user groups are involved in operational planning. There is the potential for considerable flexibility in management and for a high level of local control, subject to the ultimate authority of the District Forest Officer. Substantial forest use and harvesting are possible. In practice, many plans are not as flexible as they could be, nor do they provide as many benefits, largely because foresters find it difficult to cede control. Nevertheless, the legislation provides for flexible management and substantial benefits, and there is a significant number of cases where the agreements match the potential.

- There is no benefit-sharing by the Forest Department. At present, communities are entitled to use all products raised through management and may use all income raised for development purposes. Whether this will be extended to allow for greater levels of income from more substantial commercial use of forest products has yet to be tested.

- Increasingly, indigenous systems of forest management have been recognized. Many of these have developed in the near vacuum of forest management that existed after the nationalization of the forests in 1957. Community forest guidelines provide for agreements to be made with existing user groups. This is a major shift from the previous emphasis on official boundaries and newly established formal committees. The guidelines also permit existing groups to incorporate their management practices, where they are effective, into management plans.

*Source:* Fisher (1995)

---

**Box 3.17** *The Landcare catchment planning process, Australia*

'Preparing a catchment plan as a framework for individual property plans is a valuable strategic activity for Landcare groups. Various planning processes are evolving in different circumstances, but common ingredients include the following:

- A base map of the district is prepared, often using an enlarged aerial photograph and group members receive base maps for their own properties at a larger scale.
- The group, with the aid of a facilitator, drives and/or walks around their district, developing a common understanding of its characteristics, and agreeing on a common local language for describing the different types of land – the ecological land units.
- Group members use their local knowledge and the information generated in the group to analyse and map the land units on their own properties and this information is aggregated to compile a land unit map for the catchment.
- The group discusses land management issues and potential elements of more sustainable systems, both at the farm scale and at the catchment scale. Property and catchment planning processes can assist individual land users at the paddock and farm scales, and groups of land users at the whole catchment scale, to gather, analyse, synthesise and apply information to move towards sustainability.'

*Source:* Campbell (1994)

---

Campbell (1994) observes that:

> *'the key ingredients of Landcare are its lack of structure, the primacy of land users in determining group directions and activities, the integration of conservation and production issues, the involvement of people other than farmers in groups and the extent to which groups assume responsibility for their own problems and resources'.*

There has been a phenomenal growth in Landcare, from 350 groups in 1989 to almost 3000 in 2001, involving representatives of about a third of commercial farming operations in Australia. Some measure of this success must be attributed to a shift of government funding from state agencies to Landcare so that the local groups can commission and direct work that they believe they need. However, Martin and Woodhill (1995) point out that Landcare groups do not have structures for environmental monitoring, evaluation and regional coordination. Their local initiatives cannot deal comprehensively with issues like salinity that extend over wider areas and long timescales: the link between their bottom-up approach and equally necessary strategic planning of land use has not been forged.

The success of Landcare has been widely heralded and its influence has undoubtedly spread well beyond its agrarian roots. Lockie and Vanclay (1997) highlight struggles over the meaning of Landcare and key concepts such as *empowerment, participation, partnership* and *community*, and focus on finding suitable criteria against which to gauge its success. It is clear that Landcare is a long-term process and still faces many challenges, especially given changes in policy and funding arrangements that accompany changes to federal government. Curtis and de Lacy (1997), for example, argue that additional resources are required to increase landholder adoption of best-bet practices, and Landcare group activity needs to be integrated within regional planning processes.

Landcare has evolved in a robust democracy. Efforts are underway to examine its applicability in South Africa, but it remains to be seen whether the approach can be transferred as a generic model to other countries and, particularly, to poor countries with an immature institutional fabric.

## LIMITATIONS OF PARTICIPATION

### The quality of information

Various limitations arise from the very principles of participatory methods, particularly in relation to influencing policy. Guijt and Hinchcliffe (1998) point out that the methodologies emphasize micro-level details and diversity at the local level, local social processes and presentation in terms of specific local narratives. This means that the resulting information is too detailed for policy-makers and not often analysed in terms of policy implications. However, policy analysis can be enriched by presenting the findings of participatory approaches as case studies.

It is often difficult to quantify this context-specific information. Issues are discussed in groups and there is an emphasis on relative, rather than absolute, values. So despite the widespread uptake of participatory techniques, their findings are still greeted by the question 'but how do they compare with real data?' Pretty (1995) complains:

> *'It is commonly asserted that participatory methods constitute inquiry that is undisciplined and sloppy. It is said to involve only subjective observations and so reflect just selected members of the community. Terms like informal and qualitative are used to imply poorer quality.'*

But as Robert Chambers (1997) has reiterated, the purpose of rigour is simply to assure that data provide an accurate reflection of physical and social reality, and that personal judgement is minimized. While rigour is traditionally linked with measurement, statistical tests and replicability, these can overly simplify reality: in order to be counted, the real situation has to be dismembered. The resulting simplifications miss or misrepresent much of the complexity, diversity and dynamism of the system. Criteria of trustworthiness, authenticity, and rigour can be applied to demonstrate the soundness of participatory approaches (Guba and Lincoln, 1989; Marshall, 1990) and there can be no argument that they are technically worse or better than any other.

If the objective is data gathering, then rapidity will probably take precedence over local analytical processes. If local action is the aim, then the priority is likely to be building capacity and competence for local analytical processes. Guijt and Hinchcliffe (1998) point out that the active involvement of people and interest groups in research, analysis and planning means that all participants should have knowledge of the results. This implies effective and timely feedback, the sharing of reports, and the recognition of all contributions.

Good facilitation is the key to effective use of participatory methods, and if they are to inform policy-making and become a key part of planning, then complementarities with other disciplines must be sought.

## Costs of participation

As the value of participation in planning and decision-making has come to be accepted, some planners assume that the maximum participation of all of the people all of the time is necessary and a good thing. It is not. Complete participation may actually lead to complete inertia, due to the costs involved and practical difficulties such as transportation, reaching a quorum, time and energy (Box 3.18).

It is essential first to identify the most appropriate form of participation (whether at a local, district or higher level) that is desirable and feasible, and when particular stakeholders need to be involved. Stakeholder analysis can assist (Box 3.19). One difficulty lies in assessing a stakeholder's influence and importance. This can be investigated through public or group interviews, but it

---

**Box 3.18** *The costs of participation*

1   *Cost of providing access to information:* if people are to be actively involved in planning, they need to have a thorough understanding of the process as it unfolds and decisions that are being made. This requires effective and timely feedback, the sharing of reports and a recognition of the contribution of different groups and individuals.

2   *The cost of raising expectations:* participation may generate considerable excitement and expectations may be raised. If there is no follow-up to early discussions, disillusion may set in and jeopardize peoples' willingness to continue to participate. This can be minimized by cautious initial discussions that focus on problem identification and which provide all stakeholders with a clear idea of what is possible and what is not, given the resources that are available.

3   *The cost of facilitation:* open and frank discussions over resource allocation and use can lead to conflict that needs to be addressed. This requires specialist skills. It is questionable whether planners have these skills. They may need to be brought into a planning team.

4   *Transaction costs* of maintaining institutional mechanisms for local management include the non-market costs involved in conflict resolution, time spent in meetings, and time spent on resource management.

5   *The costs of being actively involved:* participation has costs in terms of both money and time, for local people who must take time out of already busy lives. There are also costs for food and accommodation, and the potential for political and social disputes that surface or are generated by the intervention of outsiders. These need to be compensated.

*Source:* based on IIED (1998)

**Box 3.19** *Stakeholder analysis*

Stakeholder analysis identifies the key stakeholders in an initiative (eg developing a plan or policy), assesses their interests and the ways in which these interests affect the riskiness and viability of the initiative. The *stakeholders* are the persons, groups or institutions with interests in a project or process:

- *Primary stakeholders* are those likely to be directly affected, either positively (beneficiaries) or negatively (eg those involuntarily resettled). They can be categorized according to gender, social or income classes, occupational or service use groups, and these categories may overlap in many activities (eg minor forest users and ethnic minorities).
- *Secondary stakeholders* are the intermediaries in the process (eg funding, implementing, monitoring and advocacy organizations, NGOs, private sector organizations, politicians, local leaders). Also included are groups often marginalized from decision-making processes (eg the old and the poor, women, children and itinerant groups such as pastoralists) – which may equally be considered as primary stakeholders. Some key individuals will have personal interests as well as formal institutional objectives (eg heads of departments or agencies). There may be some people who fall into both categories, as when civil servants try to acquire land in a new scheme.

Stakeholder analysis, undertaken at the beginning of a process or activity can help to:

- draw out, at an early stage, the interests of various stakeholders in relation to problems/issues which the initiative is seeking to address;
- identify conflicts of interest which will influence the riskiness of the initiative – before efforts (or funds) are committed;
- identify relations between stakeholders which can be built upon, and may enable coalitions of sponsorship, ownership and cooperation;
- judge the appropriate type of participation by different stakeholders and the role(s) each might play at successive stages of the development and implementation of an initiative.

There are several steps in stakeholder analysis:

1  Drawing up a stakeholder table – listing the stakeholders (primary and secondary) and identifying their interests (overt and hidden). Each stakeholder may have several interests in relation to the problems being addressed by the project or process.
2  Developing a matrix showing each stakeholder's importance to the success of the process and their relative power/influence (see Figure 3.5) and indicating what priority should be given to meeting their interests.
3  Identifying risks and assumptions which will affect the design and success of any actions, for instance: what is the assumed role or response of key stakeholders if a project or plan is to be successful? Are these roles plausible and realistic? What negative responses might be expected given the interests of particular stakeholders? How probable are they, and what impact would these have on the activity?
4  Identifying appropriate stakeholder participation, eg partnership in the case of stakeholders with high importance and influence, consultation or information for those with high influence but with low importance.

*Source:* adapted from ODA (1995a)

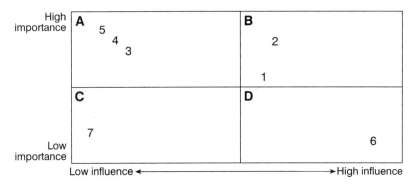

*Source:* ODA (1995b)

**Figure 3.5** *Matrix classification of stakeholders according to relative influence on, and importance to, a proposed private sector population project, Pakistan*

is often the project staff who carry out the assessment on their own. A stakeholder identification process has recently been used in the shared forest management project in Zimbabwe (supported by DFID) which led to extensive stakeholder meetings (both separately and jointly) resulting in the successful involvement of all parties in the project design process, reducing suspicion and generating a strong sense of ownership among stakeholders.

Another way to reduce transaction costs is to involve representatives of different stakeholder groups in negotiating agreements. Representativeness is a key factor for achieving durable settlements through active participation and negotiation. We develop this point further in Chapter 4, 'The resource/community level', page 167.

## Great expectations

Local participation in research and planning usually raises expectations. Consequently, before engaging in such approaches, it is important that there is a commitment to follow through with actions.

## Dealing with power

Active participation implies more multilateral relationships between stakeholders. This means that more time is needed to reach a decision. However, and perhaps more importantly, it forces shared decision-making and, thus, leads to a redistribution of power. Often, issues of power matter more than active involvement in decision-making in reaching durable agreements. Some argue that where there is significant disparity in power, and this is used to achieve an outcome, usually that outcome is potentially less stable than in situations where parity in power favours negotiation (Sidaway, 1997). We return to this issue in Chapter 4, 'Dealing with relationships and power', page 139.

## CONCLUSIONS

The last decade has seen the emergence of a new development paradigm built from the more successful examples of participatory planning and management discussed in this chapter:

- Decentralization of power and decision-making, particularly over common property resources, to local communities and user groups.
- Recognition of indigenous knowledge and practices as a sound basis on which to develop resource management systems.
- Attention paid to governance institutions, and their effectiveness, and to the role of land tenure.
- Limitation of the role of government to the creation and maintenance of an *enabling environment* that provides incentives to manage resources, rather than direction or, simply, policing and sanctions.

Yet if participation has so many intrinsic merits, why is it so difficult to institutionalize? Several constraints were identified in a study of the Indian experience of Joint Forest Management (Bass and Shah, 1994) and, although local situations will differ, many of these constraints will be present elsewhere:

- *In the initial phase, participation requires a great deal of time and effort in development of human resources.* Generally, staff are offered no incentives to make the extra effort required. Indeed, institutions and programmes are reluctant to make investments as they are currently evaluated according to tangible physical and financial targets.
- *Measurement of participation and institutional development is difficult,* requiring a combination of quantitative and qualitative performance indicators. Existing monitoring and evaluation systems cannot measure these well.
- *Participation is a long-drawn-out process and needs an initial period of interaction and evolution before being scaled-up and replicated.* Most development programmes tend to settle on a process of participation and institution-building in the early phases, without enough experimentation.
- *Participation requires a reversal in the role of external professionals,* from management to facilitation. This requires radical changes in behaviour and attitudes, which can only be gradual. It needs significant retraining to which, usually, inadequate resources are devoted.
- *Participation threatens conventional careers.* Professionals feel threatened by a loss of status if they have to deal with local people as equals and include them in decision-making. This discourages professionals from taking risks and

developing collaborative relationships with communities. National professionals feel more threatened than expatriates.

- *Programmes retain financial powers for themselves.* While many programs initiated by external agencies use participatory methods for planning, they do not make corresponding changes in resource allocation to local institutions. While it is obvious that effective auditing is essential if financial control is ceded, responsibility without financial power is just as bad as power without responsibility.
- *Participation is directly linked with equity, which threatens the existing hold of elites upon wealth, power and influence.*

Paradoxically, some of the most successful participatory projects and programmes have evolved in countries where the government has had few structures and little support for public participation. In these cases, planning structures well-suited to local circumstances have had to be re-invented, and independent groups have communicated with and worked directly with each other without being confined by the bureaucratic strait-jacket of formal institutions.

Community initiatives like Landcare and social guarantees like security of tenure cannot, in themselves, bring about sustainable land use although they may be prerequisites. They must be supported by technical knowledge, land resource information, finance (if this cannot be generated locally), good management and leadership, some of which must come from outside. Local organizations must, also, find a way to communicate their needs and proposed actions to the relevant outsiders – government and development agencies. At the same time, formal planning systems emanating from government must become more inclusive, both to reap the advantage of both technical and local knowledge and to strengthen the commitment of all parties to the implementation of plans.

There will always be development issues that are national in scope, where decisions have to be taken in the broader national interest, for example development of a major catchment for hydro-electric power or the establishment of a national park. In such cases, trade-offs must be made between national and local interests. There are other instances where the initiative should rest with local communities. Centralized, top-down methods of planning and the participatory approaches described in this chapter are not alternatives, nor are they mutually exclusive. The approach to be adopted depends on purpose and context. Clearly, links need to be forged between the two approaches so that they are mutually supporting.

Chapter 4 discusses possible ways of moving from community participation to negotiation and the development of partnerships – in effect, how to link different decision-making levels and address issues such as power structures and absorption capacity.

# 4

# A basis for collaboration

## THE NATURAL RESOURCES BATTLEFIELD

This chapter deals with the many constraints upon broader collaboration in rural development and the information needed to remove them ('Constraints and opportunities for collaboration', page 134); the different valuation of natural resources by the various stakeholders ('Valuing resources', page 146); and, finally, the institutional framework needed for participatory planning and management ('Institutional support for rural planning', page 156). There are not many examples of planning practice that have been able to bring people together to find common ground. Nevertheless, the last few years provide us with examples of attempts to manage conflict more effectively. Review and analysis of the ideas that underpin the more successful approaches provide indications of a way forward.

Participatory approaches have sometimes been presented as a panacea that will bring about harmonious and equitable negotiations. This is not born out by experience. Participation has much to offer but also brings with it a new set of problems, not least in the different interpretations placed on the term by different people, as elaborated in Chapter 3. In practice, participation in rural planning has usually been restricted to two kinds of stakeholders: community groups and development project staff. This has proved insufficient to develop sustainable initiatives in rural development, as it ignores the claims of other groups or antagonizes them. For instance, community groups may wish to continue practising traditional shifting cultivation but outside groups, such as agri-businesses or environmental groups, may be opposed – the former because shifting cultivation competes for land, the latter because they wish the forest to be left untouched. Such opposing goals often make natural resource management a confusing battlefield (see Figure 4.1).

Let us be clear that participatory planning is more than using PRA methods. It has to be associated with institutional mechanisms to ensure that local interests and priorities are, in practice, represented in negotiations. But this act of inclusion frequently upsets established patterns of power and control within a society or group. So PRA methods need to be augmented with knowledge of who should and can participate (Bliss, 1999). Innovation is required to deal with two issues. The first is to develop and use methods that allow people to collaborate, coming together to discuss and negotiate over competing claims and priorities. What are the constraints and possibilities here? The second is to provide a means of valuing the resources and associated trade-offs between differing possible ways of using a mixture of resources. Money values alone are not enough. What lessons can be learnt from past experience? We now take up these two issues in more detail.

## CONSTRAINTS AND OPPORTUNITIES FOR COLLABORATION

### Concepts and methods in collaborative management of natural resources

To build a strategy for collaboration, and before any other planning activities start, it is important to assess the likely level of competition or collaboration. Pruitt and Rubin (1986) contend that negotiation strategies depend on the way parties balance concerns for both themselves and other parties, as represented in Figure 4.2. In situations where the accommodating and avoiding strategies prevail, conflicts are likely to subside, at least in the short term.

Many would argue that collaboration is best. For example, Mike Dombeck, Director of the US Forest Service (cited in Walker and Daniels, 1997b) calls for collaborative stewardship of natural resources:

> 'More interaction between stakeholders, not only in the process (decisions and responsibilities) but also in the outcome of situations.'

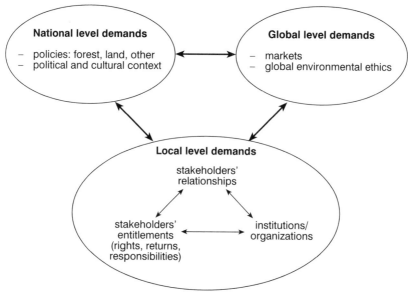

*Source:* Dubois (1998b)

**Figure 4.1** *The natural resources battlefield*

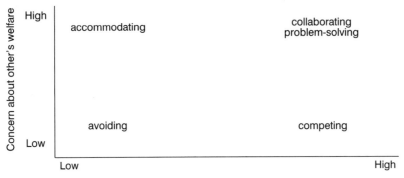

*Source:* Adapted from Pruitt and Rubin (1986)

**Figure 4.2** *The dual concern model in negotiation*

Vodoz (1994) advocates collaboration on the grounds that:

- it *improves the commitment of* stakeholders, through joint involvement in problem-solving;
- collaborative strategies are likely to *improve the overall quality of decisions*, not least through exploration of new options, with the potential for win–win settlements;
- even when they do not achieve a settlement, collaborative strategies can nevertheless *improve the quality of the failure*, because they foster discussion between stakeholders, hence improving knowledge of each other's interests and perceptions.

We would add that:

- They *allow for a freer flow of information* between stakeholders. Information flows are vital. Conflict frequently emerges because adequate information was not freely available to key stakeholders early on in the planning process (or at all).

However, collaborative strategies also have their limitations:

- Negotiation is a voluntary and non-binding process and *stakeholders can pull out at any time*, with a risk of jeopardizing the process and/or the agreement. Yet, the temptation to withdraw is often balanced by the risk of marginalization, in particular if relationships and mutual trust have improved during the negotiation process.
- They are not cost-free – high transaction costs arise mainly from the time and efforts of all parties involved in interactive participation (see 'Costs of participation', page 128, and Box 3.18). However, Bliss (1999) argues that many of the financial objections relate to an outmoded perception of the project cycle which separates planning and implementation in both theory

and practice. In other words, costs incurred early in the cycle may be abundantly repaid later and shortcuts early in the process can prove costly later.

Specialists in conflict management generally agree that collaboration is more likely to be achieved by focusing on interests rather than positions, following the *interest-based bargaining* approach (Delli Priscolli, 1997) (Box 4.1).

Three conditions are necessary for a successful negotiation (Huybens, 1994):

1 None of the parties can solve the issue alone and stakeholders have some room for manoeuvre.
2 The outcome from a joint decision is more enduring than the imposition of a unilateral solution, particularly given the complexity and uncertainty associated with natural resource situations, and the often divergent views on how to manage these.
3 There is a fair balance of bargaining power between the parties. This is a tricky issue and is discussed in 'Dealing with relationships and power', page 139. The principles in Box 4.1 place significant value on relationships (including confidence-building), power sharing, open communication and mutual gain.

## Stakeholders

It is clear that early identification of relevant stakeholders is essential for effective collaboration. Methods for stakeholder identification and analysis, describing their stakes, and agreeing how they can be incorporated into planning, are discussed in Chapter 3 (Box 3.19).

Stakeholders may view their participation in a number of ways (ODA, 1995a):

• Being in control. Only consulting, informing or manipulating other stakeholders.
• Being in partnership with one or more of the other stakeholders.
• Being consulted by other stakeholders who may have more control.
• Being informed by other stakeholders who have more control.
• Being manipulated by other stakeholders.

While nobody likes to be manipulated, it is wrong to assume that all stakeholders want to be fully in control. Some may be happy to be kept informed, or consulted as necessary. If partnership is seen to be desirable for the sustainability of activities stemming from any plan, this has to be consciously constructed in an informed manner.

Certain key issues come up again and again in developing and maintaining partnerships (Table 4.1). Some of these may be potential stumbling blocks along the path to collaboration but they can be addressed by training. Such support to stakeholders may be one of the costs of participatory planning but the returns on this investment accrue well beyond the planning period and can contribute to the sustainability of activities that emerge from the plans.

---

**Box 4.1** *General principles of interest-based bargaining*

### 1 Negotiate on interests rather than position

Positions are based on perceptions, while interests are associated with underlying needs. Conflict management should attempt to move parties away from perceptions. Once interests are identified, it often turns out that parties in the dispute have some similar interests. Reconciling interests is more likely to bring about durable settlements, based on consensus rather than a compromise. More stakeholders will be satisfied with decisions that are based on consensus than those resulting from compromise.

### 2 Separate the people from the problem, or rather, the situation

Focusing on the situation, rather than the problem, lowers expectations. This allows step-by-step progress to be made.

### 3 Emphasize progress rather than solutions

Both the environment and people's concerns are always changing. So an optimum, one-off solution is often illusory. What is needed instead is a process for constructive dialogue aiming at satisfactory agreements. Good management here equates with making progress, which leads to principle (iv).

### 4 Meet interests on substance, procedures and relationships

These dimensions are part of what Walker and Daniels (1997b) call the *progress triangle* in conflict management:

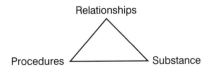

Although arguments are usually about substance and procedures, progress towards durable settlements usually hinges on relationships. However, changes in the quality of relationships are often influenced by interaction, eg through discussions on substance and procedures.

### 5 Use objective criteria in the discussion

This follows the same idea as principle (iv) – that it is more effective to discuss substance and procedures than values and emotions. Objective criteria are important when it comes to evaluating trade-offs.

### 6 Invent options for mutual gains

This implies that enough time must be given to allow for iterative analysis of the situations and interests, reaching partial agreements, searching for overall common interests among parties, etc.

**Table 4.1** *Partnerships with primary stakeholders: some key issues and ways to deal with them*

| Issue for stakeholder | Support for stakeholder |
|---|---|
| Lack of political or institutional power | Support for representative, decision-making institutions |
| Lack of appropriate information for decision-making | Ensure access to appropriate media; training. Mutual learning and sharing of available information; targeted research |
| Less powerful than other primary stakeholders | Ensure access to planning of powerless groups such as women or ethnic minorities; incorporate activities that directly benefit minorities (economically or socially) while not threatening more powerful stakeholders |
| Time and/or money costs of collaborating are high | Planning activities specifically designed to accommodate the means of all stakeholders |
| Legitimacy of one stakeholder group challenged by others | Case-by-case assessment of importance of insisting on full participation, regardless of any adverse impact this may have on participation of others |
| Discouragement by non-participatory hierarchical management structure of the agency implementing/coordinating planning process | Agency needs to change its way of working; provide help and training to adapt working practices and structures |
| A secondary stakeholder, seeking to represent interest of primary stakeholder, has management structure or value system incompatible with the primary stakeholder | Training support to NGO, or seek a more appropriate representative! |

*Source:* Based on ODA (1995a)

## Donors as stakeholders

Donors are not only a source of funds. They also have their own agendas, driven by their own political imperatives and may need quick results from the projects they support – which works against the fostering of participatory processes. In response to criticism along these lines, many now explicitly support more participatory processes with less pressure for an early definition of outputs.

However, there is still a tendency for donors to focus on sectoral issues (eg water; afforestation) rather than specifically on marginalized social groups or regions. Even where higher level objectives are framed in terms of such broad themes, they become subsumed into sectoral plans at lower (activity or project) levels (Bliss, 1999). If donor agendas are to give priority to marginalized regions and/or social groups, then initial activities must focus on a process to enable such marginalized groups or regions to develop their own plans. Planning would then become closely interwoven with the implementation of plans. This has major implications for donors: it requires that a broader set of activities (such as training and institutional support) be funded which, in turn, should support an equitable and effective planning process.

DFID policy for aid now focuses on poverty and livelihoods. This contrasts markedly with the sectoral focus that dominated only a few years ago, but DFID can draw upon significant experience to manage work in this way, for example, ODA (1995a; 1995b; 1995c). It will have to fund activities that do not result directly in physical outputs, and fit activities to rural livelihoods rather than attempting to force livelihoods into a sectoral framework. Institutional issues and, especially, the lack of institutional capacity are clearly important in any new look at planning, and these are discussed further in 'Institutional support for rural planning', page 156, and 'Better institutions to make rural planning and development work: possible ways forward', page 167.

## Dealing with relationships and power

While stakeholder relationships are important to the collaborative management of natural resources, there is not much information on ways of assessing stakeholders' relationships. GTZ (1996) categorizes relationships as:

- Service – legal/contractual;
- Market (determined by demand and supply of goods and services);
- Information exchange;
- Interpersonal;
- Power.

These various relationships are not mutually exclusive.

The interactions between relationships and power are complex. For example:

- Despite having financial and political power over local villagers, a logging company might give priority to maintaining good relationships with them, especially if it fears social unrest that could lead to the sabotaging of its equipment or the blocking of roads. It would, therefore, favour collaborative strategies.
- In Senegal, charcoal traders have a monopoly of supply to Dakar, the capital city. They give priority to maintaining their overwhelming power even if this affects their relationships with both villagers and the State. Hence, they often threaten to strike, for instance, when the government wishes to increase the tax on charcoal products. This is a situation of competition or even domination.

Figure 4.3 depicts the influence of such interactions on negotiating strategies. Two key points emerge from this analysis:

1   It is possible to reach agreements through negotiation between stakeholders, even when their positions are at variance, if the stakes/interests are less important than positions. This explains why it is important to focus on *interests* rather than *positions* in conducting collaborative negotiation, the former being more influenced by power.

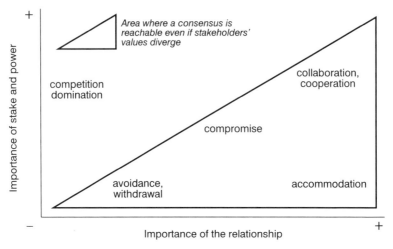

*Source:* Vodoz (1994)

**Figure 4.3** *Interactions between power and relationships in negotiation strategies*

2   Collaboration between stakeholders is usually not achievable if the importance of relationships is less important than the stakes or the importance of keeping power. Under such circumstances, negotiation should not be used to manage conflicts before bargaining powers are levelled off, as it might result in competition. Time must therefore be given to address power differences. Van Keulen and Walraven (1996) suggest some ways to deal with power differences (see Box 4.2).

Prior to negotiations, it is important to assess the power of the parties. GTZ (1996) suggests that three key issues need to be addressed:

1   *On what basis is power built?* Usually this relates to some kind of dependence: economic (eg financial dependence), social (eg hierarchical dependence, expertise) and emotional (eg personal dependence due to nepotism, cronyism, etc). Many researchers argue *that the power of the powerful is the dependence of the powerless.*
2   *How does power affect the relationship?* Power can affect the relationship physically, materially or in terms of social status. In many instances, the mere potential to exert power is sufficient to make relationships work.
3   *When and how do power relations change?* Focus on tangible elements that allow for *indirect assessment*. Vodoz (1994) suggests using *stakeholders' roles*, and assessing when and how these change. Roles, in turn, can be assessed by answering questions such as:
    •   Who has the right to do what, and how?
    •   Who does what, and when?
    •   Who is committed and willing?
    •   Who pays?

---

**Box 4.2** *Some suggested 'power-regulating' techniques*

- Use a mediator when power differences between parties are too great. The facilitator/mediator must be strong in their own right and command respect.
- The weak position should not be perceived as inherently powerless but limited by the specific circumstances.
- Start with an aspect everyone agrees upon.
- Present alternatives.
- Weaker groups should seek coalition with similarly weak parties.
- Turn objections into conditions.
- Mildly threaten. This is particularly useful when the powerful parties fear losing reputation or market shares.
- Do not allow the powerful party to lower itself.

*Source:* van Keulen and Walraven (1996)

---

- What is the best alternative to a negotiated agreement, especially for the most powerful stakeholders?
- What are the parties' means and capacities?
- What is the procedure in case agreements are breached?

Recent work by IIED on developing capacity for sustainable forestry in Africa has provided a working definition of stakeholders' roles in terms of the balance in their *Rights, Responsibilities, Returns/Revenues and Relationships* (summarized as the 4Rs). In Africa, in particular, there is an imbalance between the '4Rs' of the primary stakeholders involved in forestry. This limits local capacity to move toward more sustainable forest management:

| The State | |
|---|---|
| | • Usually has ownership rights over forest resources |
| | • Has too many responsibilities relative to its means |
| | • Often receives inadequate returns from forest resource use |
| | • Relationships with the local communities and the private sector are usually uneasy and depend on local, often covert, arrangements. There is mutual distrust among these stakeholders. |

| The Private Sector | |
|---|---|
| | • Is given concessions to exploit the resources |
| | • Is not responsible for the long-term management of resources for the public good, although it has the means |
| | • The level of returns is not clear and is a controversial issue. The private sector claims it is too low to finance sustainable forest management; yet other stakeholders believe it is high, especially when compared to the price paid for the right to exploit the resource |

- Often has opportunistic relationships with local communities

Local
communities

- Have no significant rights besides user rights. Customary rights are often more important than formal rules
- Usually have no or few formal responsibilities
- In theory, they need permits to obtain tangible benefits from the resources, and such benefits are usually small.

This situation creates an imbalance in power relationships and conflicts of interest. In turn, this makes it difficult to achieve good relationships between stakeholders and clarity concerning their roles. What prevails is a patchwork of local arrangements and quasi-open access to land resources. Use of the '4Rs' framework has helped to tease out issues and highlight leverage points in relation to collaboration between stakeholders.

Using stakeholders' roles to address power issues reduces the risk of what Vodoz (1994) calls *soft consensus*, that is a settlement where nobody formally disagrees but nobody fully backs it either, so that, there will probably not be much support when it comes to implementing whatever agreement has been reached.

## Prerequisites for collaboration

For all the apparent advantages of collaboration in the management of natural resources, it is very little in evidence, particularly as far as powerful interests are concerned. Partly, this is because of mistrust[1] between the various stakeholders, partly because the benefits of collaboration accrue to the common good and not directly to the powerful. There have been many local initiatives, some of which are detailed in Chapter 3, but their impact remains limited so long as the institutions that nominally govern rural resources lack the will, or capacity, to act as effective counterparts. There are several preconditions for improvement of this situation:

### *Political will*

Local political and traditional leaders have to accept and believe that more broadly based, interactive participation is the best path to development. Enserink (1998) goes further, arguing that participation should be a goal in its own right. He argues that (1) the insistent focus on the technical aspects of planning (solution-driven planning) means that the political discussions are ignored, only to be collided with later by a planning process that is unprepared for them; and (2) large planning projects should be considered as radical social transformation. By integrating social and political aspects into all stages of problem formulation, problem solving and decision-making, potential or actual

1 For example, Khanya-mrc (2000) cite the situation in the small rural town of Wepener in the Free State, South Africa, where the relationships between elected leaders and the residents are strained due to lack of experience, high expectations and conflicts among leaders.

conflicts will become apparent sooner. Political and social issues could become major stumbling blocks, simply because the potential for positive dialogue collapses into conflict. To avoid this, it is important to gain an understanding of the stakes of various interest groups at an early stage, so that gaps between their different perspectives can be identified and bridged.

De Graaf (1996) suggests some ways to win the necessary political commitment – including a review of formal policies and their implementation, mapping out who does what and how, and a range of mechanisms for regular interaction between decision-makers and other stakeholders (workshops, high-level steering groups, one-to-one contacts).

## Renegotiation of roles

The new development paradigm built upon participation means that various stakeholders must renegotiate their roles to accommodate the changes:

- from domination by government, private operators interests and professionals, to reconciliation of different interests;
- from management based on evidence, to a learning process that acknowledges uncertainty;
- from reliance upon technical expertise and proposals put forward by planners to the inclusion of local knowledge and proposals from stakeholders, and the premium on skills in dealing with people;
- from a narrow focus on commodities and land users, to multiple objectives including environmental management and social development.

This does not mean that all stakeholders need or want to be involved at all stages. An important part of stakeholder analysis is to ensure that those managing the planning process have a good understanding of the stakes of different interest groups, so that participation is not bogged down by an unrealistic and unnecessary pressure to get all stakeholders to participate at every stage.

## An enabling institutional environment

Participation needs institutions that facilitate rather than dictate the course of rural development. Decentralization is often presented as the solution. However, as in many other areas of rural development, there is no miracle cure here. Realism about what can be achieved is at a premium.

## Capacity

Renegotiation of roles and tasks inevitably places different demands on stakeholders but the issue of capacity development is difficult and it can be contentious. Donors (or others providing support) are often reluctant to move beyond technical matters to address factors such as management capacity and good governance, because of their political dimensions. Moreover, defining 'capacity needs' often involves value judgements to answer basic questions such as 'For what and for whom does capacity need developing?' and 'Who should assess capacity requirements?'.

## Putting stakeholder participation into practice

The greatest hurdles are the disparity in the power of different stakeholders, and mutual distrust. Space for dialogue has to be created, and enough time allowed for confidence-building measures. The case studies covering South African and Zimbabwe both recommend the need to 'institutionalize interaction'. In operational terms:

- *Directness*, through face-to-face dialogue rather than third-party association or remote communication. However, a mediator might be useful at some stage in the case of overwhelming power differences.
- *Continuity*, through repeated encounters of stakeholders over time rather than a one-off negotiation.
- *Multiplicity*, through the involvement in a range of issues rather than a single matter.
- *Parity*, with participants afforded equal opportunity for input and consideration.
- *Commonality*, through a mutually desirable objective in collaboration.

Discussion of micro-projects and stakeholders' roles has proved a useful way to begin constructive dialogue. Other ways to encourage interaction between stakeholders include task forces, joint training and workshops. Beforehand, however, there often needs to be a mechanism that increases each others' knowledge and confidence – some kind of forum for dialogue.

District Conciliation Courts in Burkina Faso and the Land Tenure Commission in Niger have both assembled a mix of local government representatives, respected members of the communities and leaders to enable a balance of opinions to be heard; but an imbalance in representation favours government representation over local representation. This means that the new bodies have struggled to maintain the semblance of independence. Tensions have also arisen from a lack of clarity about the relationship between governmental regulation and customary rules. A forum for dialogue is not a simple solution and does not necessarily lead to 'good' decision-making. Its 'rules', particularly the status of any decisions reached, need to be thought about carefully in advance and described in a way that all parties understand. If these initiatives do not work in practice, then the result is disillusion.

Various structures have been created to manage *gestion de terroir* (see 'The *gestion de terroir* approach in francophone West Africa' in Chapter 3). Bonnet (1995) has reviewed five such projects to assess the abilities of the structures to regulate and represent local interests (Box 4.3). Modes of representation are still a problematic issue and there is much experimentation to find a balance between local communities, project staff and local government. An even more serious problem is the lack of legitimacy of structures that are the creatures of the donors (who almost always shoulder the running costs). These institutions

---

**Box 4.3** *Local structure for regulating natural resource use in the Sahel: key lessons*

Bonnet (1995) reviewed five donor-funded projects in the Sahel, each of which had created structures to regulate the use of natural resources: *Projet Développement Intégré du Houetkossi et Mouhoun* (PDRI-HKM) and *Projet Développement Rural du Ganzoogrou* (PDRG) in Burkina Faso; *Projet Gestion des Terroirs Filingue* (PGTF) in Niger; *Fonds d'Investissement Local* (FIL) and *Mali Nord* in Mali. Lessons from this review include:

- The first structures were created at village level but, later, inter-village structures were introduced to widen the scope of their activities.
- The most effective bodies began informally and have evolved by taking on board early lessons from their experience of natural resource management.[2] However, this evolution is a slow process.
- Only one project (FIL-Sikasso, Mali) actually shares decision-making about the design and implementation of activities between project staff and local people. There is a tendency to subdivide the organization to separate financial and advisory functions.
- The new structures are not effective in managing local use of natural resources! Their legitimacy is limited by the power of traditional chiefs and confusion between customary and formal rules. The notable exception is the *Mali Nord* project, maybe because of the absence of effective local administration and the weakened traditional power structures following the war between the government and Tuareg groups.
- There is much experimentation to find the right balance of representation from the local communities (that is proponents of activities, village delegates and traditional chiefs), project staff and representatives from the local government agencies:
  - villagers are always present, in one project without representatives from the other stakeholders (PGTF);
  - in four of the projects, project staff are present;
  - in two projects (FIL and PDRG), villagers have refused to involve staff from local government units.
- Experience shows that, when present, the technical staff from government agencies should be in the minority, otherwise the debates tend to be dominated by technical considerations (David, 1995; Bonnet, 1995).
- The *Mali Nord* project does not even consider election of representatives, which would exclude some social groups from committees. Instead, members are appointed to reflect the existing balance of ethnic groups, gender, social groups and age according to locally decided criteria.
- A prerequisite for representation is the capacity to analyse proposals and official documents, calling for literacy.

---

depend on the goodwill of established administrations if they are to function, and they can always be undermined by political forces that cannot control them (Firmin Ouali, 1996, pers. comm.).

---

2 The choice between creating structures that answer to administrative requirements and letting them shape according to local circumstances is never easy. In the latter case, local bodies represent more local interests and are likely to last longer, but shaping takes time (eg referring to southeast Asia, Rikken (1993) considers that this process, encompassing promotion, mobilization and consolidation, may take up to 10 years).

# VALUING RESOURCES

Any decisions on the use of natural resources need to be informed by valuation of their worth. But this is not easy. It can even be difficult to distinguish high- and low-value goods. For example, fuelwood is of very high local value but of little value outside the local context (see 'Dealing with relationships and power', page 139, and Box 4.10). Clearly society can place a social value on a resource, for instance, the very high value of drinking water supplies for all members of the population, including the most poor and vulnerable. But there is also a longer-term value or environmental requirement for ensuring the sustainability of particular natural resources.

We are used to dealing with market values but many natural resources are not marketed, eg the biodiversity of a forest or its role in maintaining regional water supplies, and market values may not adequately reflect the perspective of poor and marginalized people who scarcely enter the market. Also, as indicated above, there are social values and beliefs and, indeed, societies that do not chime with the tenets of liberal economics. The following sections explore the factors to be considered and an array of ways to arrive at a robust and comparable valuation of natural resources.

## Differentiating goods and services

### Classifying goods and services according to the concepts of welfare economics

Welfare economics conceptually separates *public* and *private* goods. However, some goods are private goods in some contexts but public goods in another. Bass and Hearne (1997) discuss the concepts of *subtractability* and *excludability* to categorize goods and services:

- *Excludability* refers to the ability of an individual to deny the use of the good or service to another individual. Thus, if a good is excludable, then those who have not paid for it are excluded from consuming it.
- *Subtractability* refers to the amount that the consumption of a good or service subtracts from the repeated consumption of the good or service. In other words, if a good is subtractable, its consumption by one person reduces its availability to others.

Table 4.2 lists examples of these concepts as applied to forest goods, services and activities. Goods and services may be categorized as:

- Highly excludable and subtractable goods, like timber, are often considered to be best controlled by market mechanisms and are, therefore, considered *private goods*.
- Goods and services that are characterised by low excludability and low subtractability, such as watershed protection, do not provide much incentive for individuals or groups to replenish them, and they are therefore commonly referred to as pure public goods.

**Table 4.2** *Characteristics of forest goods, services and activities*

| Forest-based goods, services and activities | Excludability | Subtractability | Type of good |
|---|---|---|---|
| Timber | High | High | Private good |
| *Forest management services* | | | |
| Road construction | High | Medium | Depends on context |
| Fire protection | Low | Medium | Depends on context |
| Research and extension | Low | Medium | Depends on context |
| Marketing services | High | Medium | Depends on context |
| *Non-timber goods/services* | | | |
| Hunting | Medium | Medium | Club good |
| Hiking | Medium | Low | Club good |
| Camping | Medium | Medium | Club good |
| Watershed protection | Low | Low | Pure public good |
| Grazing | Medium | High | Depends on context |
| Fuelwood collection | Medium | High | Depends on context |
| NTFP | Medium | High | Depends on context |
| Biodiversity conservation | Low | Low | Public good |
| Micro-climate moderation | Low | Low | Public good |
| Carbon sequestration | Low | Low | Public good |

*Source:* adapted from Bass and Hearne (1997)

- Cornes and Sandler (1986) define a third category – *club goods* (also known as toll goods) which include goods whose benefits are excludable but partially non-rival (eg hunting). If these goods or services have significant externalities, this might justify tight control and also assistance by public bodies. However, not all members of a community benefit from club goods and services, and may resent public funds being used to support them. As a result of such sensitive situations, it is frequently necessary to make trade-offs between equity and efficiency in rural planning and management. We explore this issue further in 'Promises and realities of decentralization', page 160.
- Finally, there is the important category of the *common-pool good*: at once not easily excludable but highly subtractable, eg groundwater.

While these categories are useful in determining planning approaches and especially in establishing the roles of the various stakeholders involved in rural development, differences often depend on local circumstances and are not always clear cut. Even services with low excludability and subtractability, such as watershed protection, can sometimes be managed in a private way, eg eco-tourism activities can help to protect resources. Rather than highlight differences between categories of goods and services, it is better to view the categories as elements of a continuum with their attributes evolving over time.

Nevertheless, the differentiation between private and public goods often has been used as a key element to define planning and management strategies. For instance, governments often justify their control over natural resources on the grounds that the resources might become degraded if they fall into private hands. In some situations, the difference between private and public character is ambiguous and/or changing, and these are often better handled by partnerships between stakeholder groups. Typically, these goods and services need international support, particularly when they maintain global or regional ecosystems, as in the case of carbon sequestration.

## Differentiating agriculture from natural resources

Agriculture brings about rapid and significant environmental change over very wide areas. Some investments bear fruit within a cropping season, so are particularly attractive to the poor who cannot afford to wait for their returns. Agricultural activities are commonly in the hands of individuals or small, homogenous groups.

This contrasts with the exploitation of the natural resources which are renewable over a long period in particular forests. They are commonly exploited by powerful concessionaires and private operators are reluctant to invest in conservation or replenishment unless they receive advance financial support from public bodies. Some investments in agriculture also have relatively high costs with long payback periods and so farmers, particularly poorer ones, are reluctant to invest in them. Frequently these activities are more closely related to natural resource management than to immediate cropping needs. Soil and water conservation is a classic example of this.

Planning and management strategies for natural resources therefore require more negotiation, and often government intervention to regulate operators.

## Market value of natural capital

Collaborative management is not appealing when the commercial value of resources is low, eg when resources are so degraded that poor returns on labour make joint management activities unattractive.

When natural resources have a high market value, this generates market opportunities. This is an important factor in the success of natural resource management initiatives that involve local users. For instance, referring to tree planting, Arnold and Dewees (1995) have found that removing impediments to market access for tree products can be more effective than providing subsidies for tree planting. Another example concerns community-based wildlife management under the CAMPFIRE[3] programme in Zimbabwe (see, 'Examples of local-level resource planning', page 105) where the most successful initiatives (that is those with high revenue generation and timely distribution of revenue to households and projects) were also those with large amounts of wildlife, high species richness and low livestock and human populations. In this case, according to Campbell et al (1996), a high ratio of wildlife density to human population density explains in great part the success of the management scheme. Following this line of argument, they suggest that collaborative forest

---

3 CAMPFIRE: Communal Areas Management Programme for Indigenous Resources. See Chapter 3, 'The quality of information', page 127, for details.

---

**Box 4.4** *Commercialization pressures on common property systems*

- Incentives for appropriating the commodity and not cooperating are high.
- Enforcement of rules is likely to be complicated by high-value items, especially if the item is wanted by elites. Bribes and coercion to escape enforcement are more likely when high values bring in cash.
- Many organizations may not be flexible enough to adapt to rapid changes induced by commercialization. There may be no current rules on commercial products and there may be no past rules to learn from.
- High-value resources and commercialized products create incentives for outsiders and the state to appropriate the land and dispute legal claims.
- Legitimacy of resource use is contested by regional, national or international organizations who see their interest at stake in the use of a resource or commodity.

*Source:* McElwee (1994)

---

management initiatives in Zimbabwe are unlikely to be as successful as the CAMPFIRE project because the resource capital of the *miombo* forest is less marketable than wildlife.

Commercialization of resources also puts pressures on common property resources because of the mix of local and outside interests it generates and the power relations involved. First, there is a risk that the most powerful – local elites or outsiders – will reap most of the benefits. High stakes attract outside speculators, as evidenced by the proportion of foreign interests involved in timber exploitation in Central Africa. This inevitably increases the external pressures on local stakeholders, and even sometimes governments, while reducing their ability to control the use of the resource involved (Box 4.4).

It is difficult to reach agreement on collaborative management of highly valued resources (eg timber). In such situations, it might be best to focus first on trying to negotiate the sharing of benefits – and responsibilities and rights – for those resources in the area which have more medium-level commercial value (eg non-timber forest products). Once trust is established, and benefit-sharing is seen to be working, it might be possible to build on this and talk about sharing the more valuable resources, but this is being optimistic.

Highly priced resources also influence national natural resource management strategies. For example, in most African countries, the State asserts ownership of the trees that produce high-quality timber. Experiences of collaborative forest management are more advanced in the Sahel and dry eastern-southern Africa (with deciduous/*miombo* type forests) than in Central Africa where the monetary value of forest is higher and more important to the national economy. In the Sahel, there have been genuine moves towards decentralization of natural resource management (including forest) – though with mixed results so far; but in the Congo Basin, attempts to introduce such approaches have been no more than cosmetic.

Foreign business interests in forests in Central Africa are very important and there are great opportunities for financial gain – at both government and individual levels. Covert forest policy, driven by foreign timber companies, has significant power and influence over planning and management strategies in rural areas in that region. By contrast, in the Sahel, interests in forest resources are confined within national borders. Although these can play a significant role in local development,[4] governments themselves are not under foreign pressures and cannot generate substantial revenues from the resource base.

Some natural resource managers and academics argue that areas with poor natural capital endowment are protected from vested and powerful interests, and that this, in turn, is likely to lead to better governance and synergy between civil society and local governments. Examples of such situations have been documented mainly in Latin America where there has been a longer tradition of politically driven rural (decentralized) development than in Africa or Asia (see, for example, Tendler and Freedheim, 1994; Faguet, 1997; Bebbington et al, 1997).

## Political values

Political values have a great deal of impact on the way that negotiation and collaboration occur in practice. Political values reflect power relationships and do not necessarily coincide with market-based money values.

Conventional economic valuation, assuming market-based, apolitical forces of supply and demand determine value, ignores these political impacts on the value of natural resources. Below, we look at three major ways that political values affect decision-making.

1   *The poll factor.* Recent literature on rural planning in decentralized governmental systems shows that, almost invariably, central governments are frequently much more willing to relinquish authority over social matters (eg health, education), small infrastructure and sometimes agriculture than they are over natural resource management (eg Blair, 1997; Romeo, 1997). For instance, for several years Mali has operated a privatized health service but it is currently struggling with the development of decentralized natural resource management. Likewise, in Senegal, health and education have been decentralized since the mid-1970s but a decision to decentralize the management of forests by the end of 1996 was taken only in order to comply with the law on 'regionalization'. This reluctance to cede control of natural resources certainly has much to do with the interests at stake, as discussed above. However, it may also be related to the fact that social infrastructure and agriculture have a greater potential to secure votes, since these issues concern everybody and cut across all sections of society.
2   *The crisis situation factor.* The fact that social infrastructure and agriculture concern a broad range of citizens, might also explain why policy responses to crisis situations are more rapid in these fields than in the case of natural resources (eg IUCN-Vietnam, 1998). As a result of national and election

---

4 This role can often be negative to local development, when outside stakeholders tend to reap most benefits of resource exploitation, as in the case of the powerful charcoal lobby in Senegal.

interests, top-level decision-makers and politicians often consider that the impacts of crises concerning such matters as widespread agricultural pests or water quality must be handled more diligently than, say, crises over protected area management, soil erosion or forest degradation. Of course, this ignores the potential danger of the cumulative effects of natural resource degradation, which can have dramatic consequences, eg the recent forest fires in Indonesia and large-scale landslides in Honduras following hurricane Mitchel.

3   *The power factor.* Local power structures are of key importance in rural areas and land resources are often key components of power in such areas. Control of land tenure determines power over social affairs, be it in less-developed countries (eg the headman in African villages or the owner of *latifundia* in Latin America) or in industrialized countries (eg the lairds in Scotland). Inevitably, power disparities influence the outcomes of agreements (see, 'Dealing with relationships and power', page 139).

It is clear that the valuation methods need to acknowledge the key role of politics and power in rural planning.

## Combination of different valuation methods

Since conventional financial valuation alone is not an adequate basis for rural planning, it is necessary to combine a wide range of stakeholders' valuations.

### Combining PRA and economic methods

IIED (1994b) has reviewed the economic evaluation of forest land use options in the tropics and argues that such evaluation must encompass both production/market and non-market benefits. Three types of values are distinguished:

1   *Direct use values*, that is for consumption or sale;
2   *Indirect use values*, that is environmental functions;
3   *Non-use values*, that is cultural, religious and existence values.

Most studies concentrate on the direct use values because of the difficulties in quantitatively estimating non-marketed values. Such a bias can be misleading and there is a need to carry out a differentiated analysis. Non-economic methods such as participatory rural appraisal (PRA) can complement economic valuation methods and, also, help capture inter-annual and inter-seasonal variations in values of forest products (Tables 4.3 and 4.4).

A major benefit of combining economic and participatory valuation is that it enables the simple question 'What is it worth' to be expanded to 'What is it worth, to whom, when and in what way' (IIED, 1998).

**Table 4.3** *Key questions in valuing wildlife resources and complementarities in PRA and economic methods*

| Questions to be answered | PRA | Economic methods |
|---|---|---|
| What resources are there, and where are they ? | Participatory mapping; transects; interviews; mobility maps | Resource inventory; quantification of stocks, differentiated by cost and/or quality factor |
| When are they used/ available? | Seasonal calendars; historical time lines, and maps; product flow diagrams | Household surveys; market surveys |
| Who uses them ? | Wealth ranking; social maps | Consumption surveys |
| How are they controlled ? | Tenure maps; Venn diagrams | Analysis of market concentration |
| What are they worth – marketed values? | Household interviews; mobility maps | Household analysis |
| What are they worth – indirect use and non-use values? | Role plays; ranking and scoring matrices; daily and seasonal labour and activity calendars | Production function approach; cost-based valuation; survey-based valuation |
| How sustainable is the resource use? | Historical maps and transects; interviews with community elders; matrix ranking for abundance | Optimal control modelling; cost–benefit analysis of alternative land use options |

*Source:* IIED (1995)

## Commodity chain analysis

This approach follows the distribution of benefits from exploitation of resources lending a crucial political dimension to the analysis of the local economy. In steps:

1 Identification of the actors involved, from the extraction to the retail level.
2 Evaluation of income and profit at each level, and within each level, of the commodity chain.
3 Evaluation of the distribution of income and profit within each group along the chain.
4 Using the distribution of benefits to trace the mechanisms by which access to benefits is maintained and controlled.

As Ribot (1998a) puts it, 'Direct control over forests renders little profit. It is through control over markets that profits accrue.'

Commodity chain analysis reveals the importance of securing rights of access to the market, and some control over market conditions, as well as more secure rights to the resources.

**Table 4.4** *Limitations to economic and participatory valuation methods*

| Limitations | Effect on analysis |
| --- | --- |
| **Economic valuation methods** | |
| Concepts, terms and units imported from western experience, and are defined and interpreted in different ways by different disciplines | The definitions are critical as they structure how information is both gathered and analysed, eg the household is frequently the basic unit of analysis, but this means that intra-household and inter-household interactions, which may affect the way a particular resource is valued, are overlooked |
| Dominated by a set of assumptions that present a limited reflection of reality | Underlying assumption is that individuals and households are driven by welfare (or utility) maximization. This ignores other rational motives such as maximizing chances of survival or fulfilling social duties and rituals |
| Simplified analysis can be especially misleading in dealing with non-marketed natural resources | For example, during droughts, wild foods may mean the difference between life and death, so their value increases compared with other periods. How can these values be incorporated in long-term planning? |
| Assumes everything can be valued. Important economic role of some resources may be lost or underestimated | For example, certain species may play a vital role in rituals and so be irreplaceable. Many ecological functions are too difficult or costly to estimate reliably |
| Data collection methods can result in biases and inaccuracies (even when assumptions are relatively realistic). Interview-based questionnaires are notoriously prone to bias and inaccuracy | If data collected through questionnaires is fed uncritically into analysis, the results may be highly misleading |
| **Participatory valuation methods** | |
| Micro-level detail and local-level diversity combine with an the emphasis on social processes | Information generated is too detailed for policy-makers and is difficult to analyse for policy implications |
| Information is frequently context-specific. Relative, rather than absolute values are commonly emphasized | This makes quantification difficult as well as comparison between regions or communities |
| Inadequate facilitation skills | The dependence of the methodologies on good quality facilitation to provide trustworthy and representative findings means that, if good facilitators are absent, the data may not be of good quality |

*Source:* based on IIED (1998)

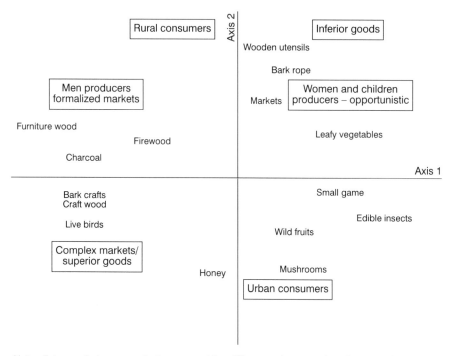

Note: distances between products represent the differences between them in terms of the 12 variables used in the analysis, eg inferior/superior good (in economic terms), gender, opportunistic/non-opportunistic markets. Only a few of those variables are shown.
*Source:* Campbell et al (1996)

**Figure 4.4** *An example of relationships among woodland products*

## Linking resources to users

In relation to forest products, Campbell et al (1996) note that literature on marketable products is scarce, uneven in its coverage (eg much on firewood, little on edible insects), often limited to classification of products, and focuses on resources rather than users. Seeking a more analytical approach, they used a multivariate technique to explore variation in product type according to criteria based both on the resource and the user, including the time of harvesting (opportunistic/non-opportunistic), ecological impact of harvesting, economic value of the products, and source of the products (see Figure 4.4). They argue that the combination of economic and behavioural approaches leads to a better understanding of: the temporal dimensions of markets and marketing channels; the contribution of natural resource products to livelihood strategies and gender differentiation; their impact on local institutional arrangements and on the status and functions of natural resources.

This knowledge can help local communities to better negotiate their rights to access, use and trade natural resources.

## Forest resource accounting

Forest resource accounting (FRA) was developed to assess forest resources but the principles apply to all natural resources. It is characterized by:

- a focus on the supply of, demand for and uses of forest resources;
- a combination of quantitative information about ecological aspects and potential of the forest asset, and more intangible elements of forest use and management, such as stakeholders' needs, roles and entitlements;
- an incremental implementation, allowing for information to be generated as needs arise and with the involvement of key stakeholders.

Figure 4.5 illustrates a typical FRA cycle and shows five process elements, one of which – agree forest goals – should be undertaken before the others. This stage requires a high degree of interaction between stakeholders, a process that itself helps build (or restore) confidence and facilitates cooperation. Indeed this is also the first essential step in the process of land use planning (see 'Putting stakeholder participation into practice', page 144).

While many places have information about standing forest resources, they do not have a system to gather and maintain information on other aspects that stakeholders agree are critical for sustainable forest management. The remaining four steps of the FRA cycle provide this information system – a feasibility study to establish information needs; if the need is established, an implementation proposal that sets out the products required to meet the needs; agreement on who should do what; and delivery of the information.

To date, national forest departments in Ecuador, Guyana, Pakistan, Himachal Pradesh (India)[5] and Cameroon have expressed an interest in FRA. In all of these countries except Cameroon, feasibility studies have been requested and undertaken between 1993 and 1998 by a team from IIED and the World Conservation Monitoring Centre. However, none of these countries has progressed to actually implementing FRA. Aside from technical difficulties in some cases, a common explanation for the delay seems to lie in the resistance to transparency and sharing of information by those who hold it (particularly government agencies) combined with institutional inertia (eg linked to very hierarchical administrations in India and Pakistan).

Clearly there is an unwillingness from the outset to share natural resources and, even, information about natural resources. Yet there is, also, broad agreement that this is the way to achieve convergence between conservation and livelihood requirements, if not sustainability. Technical and expertise-based modes of planning and resource management cannot address the diversity of interests, and arbitration often leads to less durable settlements than negotiation. The following sections consider negotiation and partnerships in the management of resources in rural areas.

---

5 In this instance, FRA has been revived recently through its inclusion as support to forest policy development (Pallot, 1999).

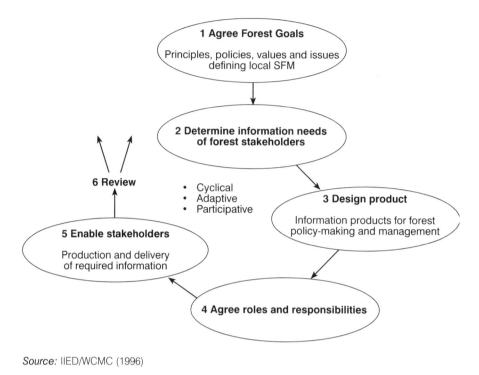

*Source:* IIED/WCMC (1996)

**Figure 4.5** *The forest resource accounting cycle*

## INSTITUTIONAL SUPPORT FOR RURAL PLANNING

### Institutional realities

The very real problems in building partnerships in rural development discussed in the section entitled 'Constraints and opportunities for collaboration', page 134, are compounded by ineffective institutions. Smoke and Romeo (1997) summarize the institutional environment as follows.

#### Complex and poorly coordinated institutional framework

The institutional setting usually combines some oversight by one central agency (eg Ministry of Home Affairs, Ministry of the Interior) over local government affairs; control of sectoral matters by line ministries (eg Agriculture, Public Works); and sometimes some involvement of a coordinating body such as the Ministry of Planning or Finance. The dispersion of responsibilities across agencies and lack of coordination at all levels are pervasive hindrances to rural development. Table 4.5 gives an example, from South Africa, of the range of systems and institutions that can be involved in planning.

#### Incentives to maintain confusion

Sectoral ministries compete for funds whether from national budget allocations

**Table 4.5** *Planning systems in South Africa*

| Planning systems | Institutional framework | Requirements | Approach and status |
|---|---|---|---|
| Land Development Objectives and integrated development planning | Local and district municipalities | Rolling 5-year plans, evaluated and adapted annually; which cover all aspects of development needs | Participatory. Involvement of authority, municipal officials and residents, in implementation |
| Water services plans | Water Services Authorities, usually TLCs | Within framework of available water resources | Top-down, expected to bring to attention of consumers and to invite comment; link with IDPs. In the process of implementation |
| Integrated transport plans | Local and district municipalities | | New legislation |
| Environmental implementation plans | Every national department with functions that may affect the environment and every province | Coordinate and harmonize environmental policies, plans, programmes and decisions; prevent unreasonable actions by provinces | Internally compiled. New legislation |
| Environmental management plans | Every national department and local government with functions involving the management of the environment | | |
| Water catchment management plans | Catchment Management Authorities | To control water use in catchment | Participatory. Users responsible for management of resource. New legislation |
| Spatial or physical planning | Municipalities, private land owners | Site and layouts. New township developments | Technical, top-down but increasingly participatory. Well established |

*Source:* Khanya-mrc (2000)

or from donors or other external sources. The more funds that individual officers have access to, the greater their prestige and the more power they can exercise. There is, therefore, little incentive for government officials to coordinate their work, as this limits opportunities for promoting projects or programmes. Poor coordination results in confusion, overlap, duplication of efforts, lack of accountability and enhanced opportunities for corruption.

## Poor public accountability

In many developing countries, this problem pervades institutions and political circles. Good rural planning requires transparency and public accountability through cost-effective means. We discuss this further under 'Promises and realities of decentralization', page 160).

## Inappropriate performance incentives

In many developing countries, promotion is linked to the ability to fulfil administrative tasks (eg to submit monthly reports, provide statistics, maintain accounts) rather than to performance in management or in promoting/ supporting development. So there is little incentive to do other than meet such administrative responsibilities. For example, foresters in Mali spend much of their effort in collecting fines, especially as a proportion of these top up the meagre salaries of civil servants. Poor performance is seldom punished and job satisfaction is low.

## Unsatisfactory donors' strategies

Donor funding often compounds the existing imbroglio by:

• adding yet another layer of poorly coordinated initiatives;
• circumventing poor local coordination by developing sectoral projects;
• channelling money into poorly equipped institutions in order to meet their expenditure schedules.

Structural adjustment programmes can also hamper rural development initiatives by shifting priorities from development to macro-economic stability and from equity to economic growth, as evidenced by the case of Zimbabwe (PlanAfric, 2000).

## Little absorption capacity

This is a controversial issue. In many developing countries, managerial and technical capacities are often weak, particularly in rural areas. However, other factors tend to foster and even sometimes create such weaknesses. For instance, a lack of financial resources increases dependency on donors and, hence, on the views and priorities of outsiders. Several examples (mainly from Latin America) show that, when given the chance and the appropriate conditions, local authorities show good skills in administering development initiatives and achieving their own objectives (for example see, Fizbein (1997) on Colombia, and Faguet (1997) and Kaimowitz et al (1998) on Bolivia). The perceived lack of capacity has been addressed by a multitude of capacity-development initiatives, but these have failed in the first place to identify weaknesses and so to assess needs. Often there are misunderstandings between the supporting (central or

---

**Box 4.5** *Aid and culture in Vietnam: sources of misunderstandings in aid programmes*

The following quotations come from interviews with government and international staff with long experience in negotiating projects in Vietnam. They provide insights into some possible sources of misunderstandings in aid programmes.

*'Courtesy demands that I am not going to disagree with you.'*

*'Many terms can be used to infer No – even Yes under some circumstances. A common way of saying no sensitively is to say "that is going to be difficult".'*

*'The government has difficulty in saying No when it disagrees with a project design element – they prefer to agree, then fail to cooperate.'*

*'To the government official, the agreement may be the beginning of negotiation; to the donors, it may represent the end.'*

*'A contract is just the legitimization of the negotiation process, not necessarily the end of it.'*

*Source:* IUCN-Vietnam (1998)

---

foreign) agency and recipient (decentralized) agencies. Frequently, these misunderstandings are caused by diverging objectives, an unrealistic scale of initiatives, conditionalities attached to the use of funds, and institutional realities such as those mentioned above. In the case of donor-supported programmes, difficulties are often compounded by cultural differences. Box 4.5 illustrates the cultural gap in the case of Vietnam.

There are also many meanings of *capacity* and donors have often ignored the political/power dimension – the capacity for good governance. Diarra (1998) emphasizes this in distinguishing between technical and institutional capacities.

*Technical capacity* focuses on the supply of skills and transfer of new technology, methods and systems. In this context, capacity development is essentially associated with training, education and technical assistance that complements local supply. Technical capacity relates to the *supply-side* approach to capacity development. A major concern is the adequacy of resources: does the country (or department, community or other organizational unit) have enough qualified and experienced staff, money, infrastructure and equipment to do the job? If not, then the implication is that missing resources should be provided. The supply-side approach has dominated aid agencies' attempts to promote capacity development.

*Institutional capacity* focuses on the ability of the country to make optimal use of the existing technical capacity and resources. The focus here is on capacity utilisation and absorptive capacity. It is more a *demand* approach (*ie* 'What are the features of this institutional environment that will encourage striving to do a good job, and make good use of the resources they have available?'). Here, the main issues are:

---

**Box 4.6** *Concepts and definitions of decentralization*

*Decentralization* is the transfer of the locus of power and decision-making either downwards (vertical decentralization) or to other units or organizations (horizontal decentralization). The power that is transferred can be political, administrative or fiscal. Five types of decentralization are commonly recognized: devolution, deconcentration, delegation, deregulation and privatization; though, in reality, most situations entail a mixture of all types. French usage is more specific: decentralization corresponds to the English devolution.

*Deconcentration* or *administrative decentralization* is the vertical decentralization of the power to act but not to decide or, ultimately, control within the administration or technical institution (eg from the Ministry of Interior to a governorship or from the national directorate of a service to the regional directorate).

*Delegation* may be vertical or horizontal transfer of limited executive, but not decision-making, authority from an administrative service to local government, parastatals or private companies.

*Deregulation* is the lifting of regulations previously imposed by a public authority.

*Devolution* or *democratic decentralization* is the transfer of power from a larger to a smaller jurisdiction, eg from national to sub-national political entities such as states or local government. This transfer may be total or partial (eg transfer to local communities of the powers needed to manage the renewable resources on their village lands).

*Privatization* is the transfer of the ownership and/or management of resources, and/or the transfer of the provision and production of goods and services, from the public sector to private entities (commercial or non-profit).

*Sources:* Goldman (1998); Thomson and Coulibally (1994); Bass and Hearne (1997)

---

- commitment of leadership;
- local ownership;
- legitimacy of institutions;
- accountability to clients;
- autonomy of organizations;
- incentives to service and improvement of performance;
- enforceability of rules.

## Promises and realities of decentralization

Decentralization is a prerequisite if local planning is to be really effective. The problem is that decentralization, like participation, means different things to different people (Box 4.6). Effectiveness in delivery of the promises requires a clear vision of decentralization, to have an agreed timetable, to plan for sufficient financial resources (and not see it as a cost-cutting exercise) and building and sustaining capacity (PlanAfric, 2000).

When using the term 'decentralization', governments often mean *administrative decentralization* – a transfer of activities within the structure of governance to local outposts without ceding power. NGOs see it as *devolution* of powers from central to more local authorities. Administrative decentralization has often been tried in response to the failure of centrally controlled rural development and service provision but has enjoyed only limited success, since problems encountered centrally are merely displaced to the local level without any increase in local accountability.

Nowadays, devolution (also called democratic decentralization) is being promoted as a panacea for local development, on the basis of institutional evidence. In Latin America, decentralization (here, also frequently called municipalization) is a hot issue, despite questions being raised over whether the process is genuinely being ignited by legitimate concerns about the constraints of centralized government, or simply as a means to masquerade structural adjustment.

Several recent reviews of the empirical evidence (Caldecott, 1996; Manor, 1997; Blair, 1997; Faguet, 1997; Goldman, 1998; Smith, 1997; Dubois, 1997) enable us to make the important distinction between the promises and existing delivery.

## Promise 1: Devolution promotes participation, representation and empowerment of marginal groups

The principal attraction of devolution is the promise that citizens will have more say in delivering the course of local development: more participation leading to better representation, leading to empowerment.

Most authors agree that devolution fosters participation in public affairs but this relates more to activities such as provision of information and consultation, taking part in meetings, signing petitions, contacting politicians and voting at local elections. It is much less concerned with establishing independent initiatives, joint decisions in planning, control of decision-making processes. Weaker sections of society participate less than more prosperous ones. Dubois (1997) and Manor (1997) identify various constraints upon community participation:

- Community members lack time and energy to invest in local politics. In Africa and southeast Asia, this is probably linked to the fact that democratic regimes have emerged relatively recently in many countries. The picture is different in Latin America, where farmers (who often call themselves 'rural workers') have engaged in a long political struggle for land rights and other entitlements.
- Community members distrust and are cynical about government authorities – central or decentralized.
- Lack of contacts between authorities and communities, due to distances between villages and administrative posts and the often large size of local authority areas.

- Even under devolution programmes, the State tends to retain a relatively high level of control on issues that are crucial for rural people, in particular rights over land and rights to earn an income from natural resources (more so in Africa than in Latin America). There is no doubt that allocating greater responsibilities without a concomitant increase in rights and income possibilities creates what Ribot (1998b) calls a *participatory burden*.
- Lack of effective representation of community interests. Candidates for election to local authorities are often members of village elites who behave as such. However, this bias is mitigated in remote areas where the population mainly comprises ethnic groups whose representatives tend to be elected alongside elites. Nevertheless, it is clear that elections alone do not guarantee good representation of local interests.
- Another problem arises where the heads of rural councils are not elected but appointed by central government – as exemplified in the case of Ghana (Botchie, 2000) and Zimbabwe (PlanAfric, 2000). This limits representation of the local population since those appointed are not accountable to the electorate, and they might favour the government rather than the local constituency in their decisions.

The effectiveness of devolution in empowering local groups depends on the groups concerned. It has been successful in the case of ethnic groups that are geographically concentrated and are able to increase their access to public debate. In contrast, women are often poorly represented on local councils and, in Karnataka in India where women occupy a third of council positions, they tend to remain silent and participate only as directed by their husbands (Blair, 1997). Similarly, mobile groups such as pastoralists may face resistance from both local authorities and more settled groups. Blair (1997) argues that, while direct empowerment of marginal groups appears difficult, indirect empowerment through advocacy and mobilization at higher levels has better prospects.

## Promise 2: Devolution entails more equitable distribution of benefits and reduces poverty

In considering the achievements of devolution, Manor (1997) distinguishes between distribution of benefits and poverty reduction. Devolution has been fairly successful in reducing financial disparities *between regions or localities*. However, this requires: (1) financial transfers from the central level downwards. This is one of many reasons why successful decentralization requires strong central governments; and (2) empowerment of poorer areas in order to effectively bargain for their interests at higher levels.

Results are less encouraging when it comes to benefit-sharing and poverty alleviation *within regions or localities*. Local elites often reap the benefits of development in rural areas, including those enjoying some sort of decentralization. Fox (1994) argues that there are two crucial prerequisites for the fair distribution of benefits and some poverty reduction at the local level: poorer groups need to be well organized and be willing to engage pragmatically with government institutions. Both conditions are met more commonly in Latin America than in Asia or Africa.

Both between and within regions or localities, improving the participation and representation of weaker sections of society has led to financial improvement of the majority of rural people.

## Promise 3: Devolution entails more financial autonomy at the local level

This is an important issue, as financial autonomy is deemed to be a key condition for effective devolution. Funds to finance local development can come either from above – through higher-level government allocation, or from within – through mobilization of local resources.

### Central government allocation to local government units

Manor (1997) argues that allocation from central government is necessary in the first stages of devolution, while local government units (LGUs) build credibility among their constituencies. Financial transfers from central government are also necessary to overcome disparities between regions, with poorer ones less able to generate their own funds. Blair (1997) reviewed cases from Bolivia, Honduras, Karnataka (India), Mali, the Philippines and Ukraine, many of which revealed generous central government allocations to LGUs. But allocation from the centre usually comes with strings attached. This constitutes an indirect form of control of local affairs by central government although there are examples where LGUs have proved quite capable of administering local funds when given sufficient autonomy, notably in Latin America (eg Fizbein (1997) for Colombia; and Faguet (1997) for Bolivia). Faguet (1997) in Bolivia found that funds are administered better in poorer municipalities than in wealthier areas, maybe because corrupt politicians tend to seek office in relatively resource-rich areas, and the social means of supervision and control are stronger in small rural areas. This is linked to the issue of empowerment in benefit-distribution discussed in the previous section.

### Mobilization of local resources

Local resources may be mobilized through local taxes, or through investments in cash or kind by local people. Tax collection following decentralization has usually proved unrewarding, not merely because of scarcity of money in many rural areas but due to more informal factors (Blair, 1989):

- At central level, a reluctance to equip LGUs with authority and the means to raise local taxes.
- At LGU level, a disinclination of local politicians to annoy their constituents – especially the people who own or control most of the locally valuable resources but who are, also, those in the best position to evade taxes. Moreover, LGUs often lack managerial skills to handle local taxes and also financial resources to collect them.

- At community level, an unwillingness of local residents to pay taxes, dismayed by failed previous experiments with decentralization and overall distrust of government officials;
- Local communities may not administer their own share of local taxes equitably, especially where they lack social cohesion.

One way to promote *local investment* is to transfer functions from the government to the market. *Private* and *club* goods and services (see 'Differentiating goods and services', page 146) – which have higher market value – are the most amenable to private management. However this may not be appropriate for multiple objective (conservation and development) projects and can easily further marginalize the poor. Co-financing through the State or donors may mitigate these problems.

Another way to foster local investment is through stakeholders' associations and partnerships. Such associations are typically concerned with club goods and services that are influenced by economies of scale, eg rental of equipment or acquisition of technical information by farmers' or pitsawyers' groups. On the other hand, partnerships are often cited as powerful mechanisms for developing local capacities, including in investment (eg Fizbein (1997) cites a study of Argentina, Bolivia, Colombia, El Salvador, Jamaica and Venezuela; WCFSD (1999) gives examples for Africa; and Thirion (1997) discusses the development of partnerships in Western Europe under the EC-funded LEADER Programme). However, willingness to engage in partnerships and their quality depends largely on the type of stakeholder relationships and power structures at local level as discussed in the section entitled 'Putting stakeholder participation into practice', page 144.

Experience shows that local mobilization of resources has to be complemented by allocation of funds by central government so, despite what many proponents of decentralization claim, it may actually lead to an increase in overall government expenditure. However, the constraints discussed above are overcome as funds are likely to be used more cost-effectively than in highly centralized systems.

## Promise 4: Devolution improves local accountability

Accountability mechanisms are the crucial element in successful decentralization. Kullenberg et al (1997) identify three types of accountability: accountability of civil servants to local leaders;[6] accountability of local leaders to local citizenry; accountability within decision-making bodies.

1   *Accountability of civil servants to local leaders* has been difficult to ensure, mainly because of the incomplete devolution of authority from central to local authorities. Blair (1997) gives the example of Bolivia where a mayor can order his health officers to keep required clinic hours, but has no authority

---

6 Given the shortcomings of elections in many developing countries – already alluded to – we prefer to use the term local leaders rather than elected officials.

to sanction them if they fail to do so. Also, local civil servants are often reluctant to be placed under the authority of local officials because the prospect of a future post at central level is a greater incentive than being acknowledged for performing well in contributing to local development. Improvement will come only if the structure of incentives to perform well is reversed, and if high-level authorities support local ones to achieve this.

2   *Accountability of local leaders to local citizenry* might be improved by devolution through the role of elections (eg Blair, 1997; Manor, 1997), but elections do not guarantee representation of marginal groups. Based on experience in Africa, Anyang' Nyong'o (1997) declaims the 'fallacy of electoralism'. He argues that democracy needs not only free elections but also a wide variety of other mechanisms to represent civil society and guarantee checks and balances in government. These include 'independent legislative and judicial bodies, interest groups, civic associations and political parties within society'. Under 'The local government level', page 172, we discuss further various ways to increase accountability to citizens.

3   *Accountability within decision-making bodies* – two types of entities are considered here: government line agencies and civil society associations. Accountability within line agencies is critical with respect to financial transfers from central level. It also concerns the need for local authorities to comply with higher-level strategies: lack of accountability to higher levels often leads to contradiction between local and national rules. For example, in Vietnam, provincial and district authorities raise taxes on land tenure certificates although this is forbidden by national decree. Accountability to higher administrative levels is necessary to moderate excessive administrative autonomy. Lack of accountability within village-based associations and NGOs often leads to all sorts of abuses. Development initiatives can provide opportunities for local despotism leading, in particular, to the unfair distribution of benefits and the exacerbation of existing power disparities.

## Promise 5: Devolution increases the effectiveness of LGUs in delivering goods and services

Devolution usually improves service delivery by local governments, in particular where governance is enhanced and where LGUs have increased financial and administrative autonomy. However, local development is often biased towards small-scale social and economic infrastructure. This is, of course, welcome as it improves the living conditions of the majority of rural people but this bias ignores the calls for action in the area of natural resources (which are often prioritized during participatory exercises at community level). This highlights the frequent lack of communication between communities and the lowest administrative levels, even under devolution programmes.

The relationship between devolution and the quality of government services also depends on the size of the area in question. For small countries, such as many islands in the Caribbean and the Pacific, regional cooperation is preferable to decentralization in dealing with regional environmental matters.

In the field of planning there are some weaknesses in the performance of local government when compared to higher administrative levels (Manor, 1997; Blair, 1997; Romeo, 1998; Kullenberg et al, 1997):

- There are tensions between local planning and local politics. It is very difficult to change the attitudes of authorities and planners in favour of more active participation, especially when the local planners are selected from local administrations. Local authorities tend to resist formalized interactive participation as it reduces their own discretion. Such resistance is not surprising when there are no incentives to enhance synergy between State and civil society (eg where participation is not seen as a means to prevent conflicts, or where recognition is not linked to service delivery) and where money is not available to finance participatory processes.
- Local capacity is often lacking. This problem may be exacerbated by the surge in demand for tangible achievements engendered by participation.
- If priorities are not set, resources become spread too thinly in order to provide improvements everywhere. Kullenberg et al (1997) suggest that equity considerations should be balanced with concern for efficiency, though this would be hard to justify locally.
- Local elites will try to take advantage of profit-making opportunities opened up by local planning.
- Elected leaders often see planning as the compilation of wish lists during the annual budgeting process.
- Decentralized planning can lead to bureaucratic inflation which overburdens already limited local capacities. Blueprint development initiatives often have the same impact. Many donor-supported rural development projects call for the development of long-term, village-based natural resource management plans; but these are often unrelated to the daily coping strategies of rural communities, and the plans are mainly used for showing to outside visitors. We discuss the linked issue of subsidiarity and ways to improve local planning in the following section entitled 'Better institutions to make rural planning and development work: possible ways forward'.

Decentralization has rarely lived up to expectations because it has been deconcentration, not devolution, and has not addressed what Richards et al (1996) call the *invisible institutions problem* (individuals seeking financial gain from assets they control – but do not own – patronage, personal power struggles and negative attitudes to participation). More recent attempts to introduce devolution show more promise, especially in socially homogenous areas with poor natural resources. Priority has been given to the easier tasks: social matters and providing small-scale infrastructure for education and health, rather than income-generating activities and the management of natural resources.

Examples of successful local initiatives that *do* include the management of natural resources are only islands of success and lessons are rarely fed into the wider process of development. One reason for this is because natural resource management requires that politically sensitive issues are addressed (eg land tenure, control over resources) and this can be seen as threatening by local and national elites. In short, while decentralization might improve local management, it is not a prerequisite, nor a guarantee of good local management.

To be effective, decentralized systems must have:

• enough power to exercise substantial influence over political affairs and over development activities;
• sufficient financial resources to accomplish important tasks;
• adequate capacity (both technical and institutional)[7] to accomplish those tasks;
• reliable accountability mechanisms.

Two factors seem to be key in designing support programmes to meet these requirements: (1) they should be tailored to the local context rather than desired outcomes and imported principles; (2) they need to acknowledge the highly political dimension of local development, and deal with the *invisible institutions problem* in a pragmatic and non-antagonistic fashion.

## BETTER INSTITUTIONS TO MAKE RURAL PLANNING AND DEVELOPMENT WORK: POSSIBLE WAYS FORWARD

In this section, we focus on some possible ways to improve the management of natural resources, planning and development. Three levels of management are discussed: the resource/community; the local government and the central State; and approaches to linking these different levels.

### The resource/community level

At the resource or community level, we are concerned only with the involvement of communities and private operators; government involvement is dealt with in the next section. Local communities are not used to having formal authority to make their own decisions on natural resource management. More usually, government officers tell local people what is to be done. Community members tend to be reluctant to accept responsibilities for resource management, and community management will be difficult to achieve if the proper ingredients are not in place. These include:

• real power and rights – if not ownership rights, at least management rights allowing villagers to commercialize resources without needing to follow cumbersome and sometimes restrictive procedures;
• competence;
• economic interest.

---

7 See 'Little absorption capacity', page 158, for definitions of technical and institutional capacities.

---

**Box 4.7** *Success criteria for community institutions managing local natural resources*

The chances of success are greater when:

1    The costs of exclusion (eg fencing) are high, but only if local stakeholders are paying themselves.
2    The rules for access to the resources make sense in terms of the local economy.
3    Relationships between resources and user groups:
     •    there is overlap between where users live and the location of common-pool resources,
     •    the resource is vital to users,
     •    users know and understand sustainable management practices.
4    User groups:
     •    groups are small – no more than 30–40 members,
     •    boundaries are clearly defined,
     •    local resource users are able to maintain their rights to the use and management of common resources in the face of competition from both outside and within the group,
     •    there are good arrangements for discussing common problems,
     •    the mutual obligations of users affect their social reputation,
     •    there are rules, and graduated sanctions to punish offenders,
     •    there is consensus about who the users are, and this is established when the group is established.
5    Most individuals affected by the rules can participate in modifying these rules.
6    Monitors, who check and report on compliance with the rules, are themselves resource users or are accountable to the users.
7    All parties have rapid access to low-cost conflict resolution mechanisms.
8    The rights of users to devise their own institutions are not challenged by external governmental authorities.
9    Governance and provision of services are organized in a nested hierarchy, matched to the geographic scale of the resources to be managed – very local for management of village wells and pastures, large scale for catchment water management, larger still for integrated coastal area management.

*Source:* Hobley (1995) after Wade (1988) and Ostrom (1990)

---

The '4Rs' framework, introduced in 'Dealing with relationships and power', page 139, is one way to address these issues, not just for communities but for all stakeholders involved in natural resource management. Box 4.7 sets out criteria to assess the probable effectiveness of institutions and organizations involved in local natural resource management.

People's participation is not enough to bring about sustainable management. There has to be adequate representation in decision-making bodies and empowerment to ensure local communities have bargaining power in the negotiation over local resources. Ribot (1998b) argues that *representation* is crucial for it addresses the issue of who shall have control over resources and benefits. If communities do not have adequate representation, participation is pointless because they cannot (as a whole) interact meaningfully with other stakeholders.

---

**Box 4.8** *Re-introducing traditional leadership in Zimbabwe*

Zimbabwe plans to establish village and ward assemblies which will re-introduce traditional boundaries and traditional leadership and allow a voice to all villagers. These may be contradictory aims, but the move illustrates the desire to test new forms of community-level institutions, recognizing that the old administrative structures of VIDCO and WADCO (see Box 3.9) have largely failed. However, the costs are huge and will be likely to delay the implementation of this plan for a considerable time. Several thousand villages and wards are involved, each of which will require definition, assistance in being established, capacity-building. Under the new Traditional Leaders Act, villages are now seen as the guardians of the land and natural resources. The outstanding questions are: Where does Zimbabwe find the financial resources to implement the Act and how can democratic and traditional leadership principles be balanced?

Under the proposed national Land Policy Framework, the accent is on village democracy, with traditional leaders placed in honorary positions. This is probably not what the government had in mind.

*Source:* Derek Gunby, PlanAfric (pers. comm.)

---

Mechanisms that complement local elections and which can enhance the representation and empowerment of communities include:

- Mandatory inclusion of members from marginal groups in local councils, for instance women in the Philippines and Mali, the scheduled castes in Karnataka.
- Ensuring that a proportion of the voting membership of local councils are members or representatives of NGOs (eg 25 per cent in the Philippines).
- Local mechanisms that promote trustworthiness:
  - Setting up structures in parallel to the local administration, where important decisions are actually taken. These can be formal, like the local councils in Uganda, or informal and more culturally entrenched, as in Tanzania and many places in francophone West Africa.
  - Working through traditional leadership structures, eg the re-introduction of traditional leaders and traditional leader boundaries in Zimbabwe (see Box 4.8).
  - Creating sub-committees in order to separate management from advisory functions, eg in some *gestion de terroir* projects.
  - Appointing *honourable citizens*, who command respect in local parastatal conciliation bodies, such as district conciliation courts in Burkina Faso (Box 4.9), village forest committees in the Duru-Haitemba and Mgori woodlands in Tanzania (Box 4.10).
  - Allowing independent candidates to stand for elections to decision-making bodies. In extreme cases, political militants have been barred from standing in elections, eg in Burkina Faso, Uganda and Ghana.

---

**Box 4.9** *District Conciliation Courts in Burkina Faso*

District Conciliation Courts (DCCs) have been used since 1993 to settle land disputes. They are chaired by *Préfets* (District Officers) and composed of four lay-assessors. These are neither customary chiefs, nor political militants, but rather honourable citizens from the community, eg retired school teachers, and former civil servants and, to a lesser extent, younger high school graduates who usually play the role of secretaries. The blend of a central government representative and respected community members provides a balance of State interests and community voice, hence a fair representation of interests.

However, the DCCs and the *Préfets* have been facing problems in applying rules, be they inspired by community norms or formal law – often due to the co-existence of formal regulations and customary rules. In practice, it has resulted in:

- A dichotomy in which DCCs and the *Préfets* have either total control or no control at all over the outcomes of land disputes. This usually stems from either rigorous application of State regulation, or laissez-faire, leaving decisions to local groups.
- In some instances, State coercion to settle conflicts over land.

The regulation of land use and, by extension, of natural resources, seems to combine the logic of State law (aiming mainly at achieving order) and the substantial, although sometimes radical, logic of local communities, often leading to new local laws.

*Source:* Lund (1996)

---

Another approach to empowering community-level groups is the development of alliances – often with NGO assistance; which helps in establishing credibility vis à vis other stakeholders and, in particular, the State. Box 4.11 provides an example of a local association of small private operators, created without any outside assistance.

Regulations and institutions to control the use of resources by small-scale entrepreneurs are often weak, but, in some cases, their activities are controlled and/or legalized, eg:

- In some instances, there are informal arrangements between individuals: for example, in some areas of Zambia, charcoal burners have informal arrangements to help shifting cultivators clear the land and, in return, they benefit from the felled trees (Makano et al, 1997).
- In other instances, communities themselves exert a regulatory role, as in the case of rural markets in Niger (Box 4.12).
- In the Ivory Coast, many technical tasks have been subcontracted to local entrepreneurs and/or forest cooperatives by SODEFOR, the parastatal body in charge of forest management, and generate significant off-farm income at local level.
- Through the development of associations (see, for example, Box 4.11).

**Box 4.10** *Local forest management in the Duru-Haitemba and Mgori* miombo *woodlands, Tanzania*

Up to 1994, these woodlands were under government management. Their condition has been steadily degraded by encroachment for farming, grazing, hunting and charcoal making by local people, and timber extraction, mainly by outsiders.

In the case of Duru-Haitemba forest, plans to establish a forest reserve prompted local inhabitants from five villages to take as much as they could as fast as possible. Hence, local authorities informally agreed with project staff that inclusion in the government gazette would be suspended, subject to demonstration that local communities could halt forest degradation. This implied that local communities would be responsible for daily management and control of the forest, and launched a very dynamic process by the villagers, including:

- Drawing up simple but effective management plans which, mainly, comprise rules concerning the use of the resources, and the modalities of fining offenders. Any use considered damaging was banned, even if it provided income to some local groups (eg charcoal burning, tree felling). Plans were refined and validated during village assemblies.
- Setting up of village forest committees, as decision-making bodies in parallel with the official village councils. The composition of the committees rapidly shifted from village leaders to ordinary villagers, as a response to previous mismanagement of funds and the consequent need for greater accountability.
- Selection of 100 village forest guards to control abuses by both outsiders and insiders – the government had previously employed only two forest rangers.
- Subsequently, village plans and rules have been re-written as bylaws, formally approved by the district authorities to provide the necessary legal backing to informally derived regulatory instruments. This and the entitlement of village land have turned communities into legal owners and managers of their own forest reserves.
- Local foresters and outside advisers have provided only *ad hoc* assistance, eg advice on legal matters and resolution of inter-village boundary disputes.

Since the process began, degradation of the forest has reduced significantly, at no cost to the government. On the other hand, communities have gained power, through ownership, increased responsibility but also access to benefits from the resource. Finally, local foresters have been liberated from the coercive role that failed to protect the forest, and are better able to provide technical assistance.

In 1995, a similar dynamic process was launched by local communities in the Mgori forest, with similar outcomes.

A major reason for the success to date of this process lies in the fact that it has been locally initiated, without outside pressures, nor the use of blueprint schemes. The important role of the government has been to let go of its powers.

*Source:* Wily (1996)

---

**Box 4.11** *The case of the Masindi Pitsawyers and Wood Users Association (MPWUA) in Uganda*

MPWUA was legally constituted in 1994 and currently has a membership exceeding 100 pitsawyers. The association requested a concession in a local forest reserve in exchange for helping to control illegal activities.

Bylaws developed by MPWUA are aimed at preserving forest resources. For instance, each pitsawyer cannot have more than four saws. Moreover, MPWUA has a say in the issuing of sawing permits by the District Forestry Office.

The main incentives for the establishment of MPWUA were:

- Financial: secured revenue, as legally processed wood will not be seized and is, therefore, guaranteed to provide income.
- Environmental: assisting the forestry service in counteracting illegal pitsawing and other wood uses, contributes to sustaining their source of income.
- Social: achievement of good relationships with the local population and the forest department staff.
- Technical: as a group, it is easier to justify and finance training, eg for forest management, waste recovery, and other types of assistance. In 1995, MPWUA funded the Forest Department to map timber stocks in the concession area and it plans to help in opening and maintaining local roads in the concession.

The activities of MPWUA are also advantageous to the Forest Department:

- they assist in the control of illegal activities;
- they have increased the collection of taxes from pitsawing and other wood uses;
- MPWUA provides financial assistance for forest management tasks.

*Source:* MPWUA Chairman, pers. comm., 1997

---

# The local government level

## Assessing local institutional capacity

The local district government is the local provider of public services and rural development planning. At this level there may also be structures outside of government, some with a narrow remit (eg internal drainage board, hospital management committee) and some with a wider remit (eg Landcare groups in Australia, inter-village committees in *gestion de terroir* projects in West Africa). Existing institutions at this level are hugely variable in size, quality and capacity. Prior to further capacity development for rural planning, the existing capability should be assessed. Kullenberg et al (1997) have suggested a framework to assess local institutions in terms of what they call *institutional topography* and political commitment to transfer power and resources to the local level.

The local institutional topography relates to tangible features of the areas under the authority of local authorities, which have clear implications for the scope of local development programmes. To illustrate this, Kullenberg et al (1997) discuss the contrast between: (1) Mali, where *Communes* of 10,000–15,000 inhabitants have a few low-grade employees, a budget of several thousand

---

**Box 4.12** *The rural markets in Niger*

In the 1980s, Niger's firewood marketing system, based on government control through permits, was anarchic and mainly benefited merchant-transporters. Donors promoted reform under the *Energie II* project which culminated in 1992 in a new law that introduced radical changes:

- The creation of rural markets (RMs), subject to the existence of local management structures (LMSs) composed only of representatives who have usufruct in the area. In practice LMSs are created at village level and represent different user groups (woodcutters, farmers, herders; but women have not been included). The Forestry Service chooses the villages according to production potential. LMSs manage the RMs and supply them with woodfuel.
- Three types of markets have been created:
  - *Controlled* markets, supplied by delineated and managed production zones;
  - *Oriented* markets, supplied by delineated but non-managed areas;
  - *Uncontrolled* markets, tolerated during a transition period.
- Taxes are now based on volume of wood transported to the cities. Firewood from RMs is charged less than that from other areas as an incentive for traders to purchase where production is organized by villagers. Tax revenues are to be divided between the Public Treasury, the LMS and the local municipality. The more controlled the market, the more revenues are allocated to the LMS. Tax recovery within RMs has been almost 100 per cent.
- Annual quotas are determined by committees comprising one representative of the LMS, two Forestry Service officers and one municipality staff member.

Between 1992 and 1995, 85 RMs were created covering an area of about 352,000 ha. In 1995, they supplied 15 per cent of the needs of Niamey. The creation of new RMs has been steadily increasing – mostly in the oriented markets.

A national information campaign aimed at rural populations, merchant-transporters and urban consumers was seen as a prerequisite to the launching of activities, to reduce rural people's distrust of the official dialogue promoted by the 1992 Law.

Despite its achievements, the project has faced some problems and several criticisms have been made concerning:

- the overemphasis on woodfuel, at the expense of other categories of forest resources;
- the exclusion of women from the LMS;
- the slowness of the process of registration of LMS and RMs – actually, the process is being privatized, with consulting firms and NGOs assisting the local forestry service in the follow-up process;
- the merchant-transporters lobby putting pressure on the government to overturn the 1992 law and resume the old uncontrolled trade – this has been resisted by the government;
- its dependence on cities as important market outlets, which limits its scope.

*Source:* various authors cited by Dubois (1997)

---

dollars, but no fiscal autonomy, since all locally generated revenues must be sent to the central treasury; and (2) Uganda, where District Councils govern at least 500,000 people, have several hundred employees (many graduates), a budget of several million dollars, and clear statutory fiscal and legislative powers and responsibilities.

The mapping of local institutional topography in the form of simplified databases also helps to make planning decisions more objective, offsetting the *ad hoc* allocation of funds on political grounds (Smoke and Romeo, 1997).

Political commitment for LGUs to transfer power and resources to the local level has already been discussed in the sections 'Prerequisites for collaboration', page 142, and 'Putting stakeholder participation into practice', page 144. Development institutions outside government are usually recognized at national level only if they serve the interests of the government. The fact that committees established by donor agencies in *gestion de terroir* projects are not recognized by government severely hampers their ability to manage natural resources. In contrast, the adoption of spontaneous Landcare groups and the financial support from the federal government of Australia has lead to an impressive expansion of their area and scope of operations. The formalization of self-governance for various bodies can prevent their suspension by central authorities on a spurious pretext. It makes sense to legitimize decision-making structures – formal or informal – where they achieve good results.

## Autonomy to undertake development activities and modify local rules and institutions

LGUs require this autonomy in order to secure support from local political leaders and local public interest. To achieve this, locally elected bodies also need sufficient resources (notably financial) which can be provided initially from higher levels, as long as mechanisms are introduced to progressively enable self-financing. There is the usual risk that local elites will use increased autonomy to further their own interests but Olowu (1990) suggests that this risk can be reduced by providing autonomy in an incremental fashion, linking it to performance.

The United Nations Capital Development Fund (UNCDF) has developed a pilot programme on Local Development Funds (LDFs) to provide financial autonomy to local authorities (Box 4.13).

## Greater accountability of local institutions

In designing development programmes, the provision of checks and balances to guarantee fair distribution of profits from local use of resources is often overlooked. The State has an important role to play in this, both at central and local levels. One of its roles is to hold regular and fair elections. However, where a government's ability to regulate local affairs is weak, other mechanisms can assist in promoting accountability, for instance:

• *Improvement of citizens' access to information*, thus enabling more informed participation in public debates. This can be achieved through the use of local media (eg through the hundreds of AM broadcasting stations in rural areas of the Philippines); and training in numeracy and literacy.

---

**Box 4.13** *The Local Development Fund programme of UNCDF*

Over the last seven years, the Local Development Fund (LDF) component of UNCDF has developed programmes that provide capital budgets and technical support to local governments and decentralized State authorities in various less-developed countries. These projects average US$3–9 million over 3–5 years, typically corresponding to $1–5 per capita. A key aspect is participatory planning and building capacity at local government level to develop viable development activities. Important features of LDF projects include:

- In contrast to most other types of programmes supporting local financial autonomy, allocation of LDF funds is not demand-driven, as this often results in funding wish lists. Rather, ceiling funds are allocated to match existing transfers from central level.
- The funds are fixed to force local authorities to prioritize actions. Participatory planning is used as a tool to facilitate prioritization.
- An up-front entitlement is provided to promote the mobilization of local funds.
- LDF projects focus on local governments because they are assumed to have a comparative advantage over NGOs in delivering a range of infrastructure and economic development.

LDF projects typically face three key challenges:

1   Ensuring the transparent allocation of resources.
2   Making planning participatory. The planning process is entrusted to a body that must be representative both of local government and civil society. Some planned activities may be beyond the scope of LDF and local authorities, eg private income-generating activities or *common-pool* degraded natural resources.
3   Linking activities to natural resource management. Given the focus on local governments, LDF activities tend to be biased towards small-scale and social infrastructure.

*Sources:* Kullenberg et al (1997); Smoke and Romeo (1997); Romeo (1998)

---

- *Mechanisms to control daily operations* which are based on shared responsibility, eg the need for several signatures to approve financial expenditure.
- *Transparency* in review and authorization of contracts and verifying expenditures.
- *Formal procedures of redress* against elected officials. This is essential for the mobilization of local initiatives in the long term. But such mechanisms, where they exist, are often deliberately designed to be cumbersome so as to limit their use by local people. Recent innovations in Bolivia which aim to overcome such problems include a new law that sets special circumstances under which municipal councils can formally dismiss a mayor, and the creation of municipal vigilance committees (see Box 4.14).
- Better *representation* of local interests.

---

**Box 4.14** *Municipal vigilance committees in Bolivia*

These committees were established following the Popular Participation Law of 1994, to oversee municipal spending and propose new projects. They comprise representatives from local groups within each municipality, and are legally distinct from local councils. At first, local people opposed the creation of such committees, fearing that they would compete with their traditional unions (*sindicatos*). However, they changed their minds when they realized that these groups could be part of the vigilance committees.

The power of the vigilance committees lies in:

- Their ability to lodge an official complaint to the senate seeking to suspend disbursements from central government to a municipal council if it is judged that such funds are being mismanaged.
- The moral authority that they command at local level, which stems from their degree of representation of citizens' interests.

*Sources:* Faguet (1997); Blair (1997); Kaimowitz et al (1998).

---

## Subsidiarity

The principle of subsidiarity is the devolution of powers to the most local level of government that can effectively discharge these powers. In the case of rural development, this principle may be interpreted in various ways.

First, in many developing countries, subsidiarity often happens by default when central government is so weak that it exerts little control in rural areas. Then, local management of natural resources is a patchwork of local arrangements that tend to favour the most powerful stakeholders – a situation that is usually not conducive to sustainable development.

Secondly, for the sake of economic efficiency; services should be controlled and financed at that scale where there is no overlap and decentralization of management should concern goods and services for which economies of scale cannot be achieved.

Thirdly, while economic efficiency considerations are important, so is social efficiency. Weber (pers. comm.) argues that devolution should be linked to the highest level at which stakeholders know each other well enough to be able to control each other in a cost-effective way. Equity may also come into play. As discussed earlier, decentralization may lead to greater inequities between social groups if accountability and representation mechanisms are weak.

Finally, there is the crucial and contentious issue of capacity, particularly that related to governance (eg transparency of information, accountability). The debate here often revolves around the tension between long-term support to build capacity and the short-term efforts of donors to meet their disbursement schedules while achieving tangible results to justify the use of the funds.

Debate continues about the appropriate roles and levels of devolution, especially over the choice between communities and LGUs. The appropriate level of subsidiarity might be negotiated with local stakeholders and will vary according to the resource at stake. It is sometimes argued that local governance at village level is not cost-effective, but the costs of governance can be reduced by simplifying procedures and conditionalities associated with the local management of natural resources. For instance, land use management plans at village level could be replaced by locally decided sets of norms and rules,[8] as in the case of the Duru-Haitemba and Mgori woodlands in Tanzania (Box 4.8).

Table 4.6 shows that the appropriate level of subsidiarity depends on the resource/sector concerned. It also demonstrates that, apart from issues which have an obvious national scope such as land reform, the administrative level immediately above villages is often key in decentralization – it is the first level where marketing operations occur and where the State should begin to play its role as a referee. This explains why planning initiatives, such as UNDP's Local Development Fund Programme (Box 4.13), focus on the district level. In Zimbabwe, the Give a Dam campaign shows that field operations should be based at the district level, although the programme itself is coordinated at the provincial level (Box 4.15).

Linkages between different levels of authority must be developed to avoid inconsistency between initiatives managed at different levels. This has proved to be a challenge. The issues fall in two groups: linking between different administrative levels; and linking communities and the lowest administrative levels.

*Linking between different administrative levels*
This linkage requires incentives for cooperation between different administrative bodies at a given level in order to take advantage of economies of scale. This will have to be fostered from higher levels, eg through the establishment of consultative platforms. But how to finance these platforms? In many instances, donor support is needed. Another approach is to separate formal coordination functions from day-to-day collaboration (Dubois et al, 1996).

A further challenge is to promote the involvement of local authorities and communities in the planning, financing, implementation and monitoring of regional/national programmes. Contractual arrangements might be a way forward here (Romeo, 1998).

*Linking communities and the lowest administrative levels*
This benefits planning by garnering local initiatives on natural resource management. Local administrations also play a key role in providing support to community projects and formalizing of village entitlements. The federating of different interest groups in communities has proved helpful in coordinating their collaboration with LGUs. NGOs often help communities voice their

---

8 Local management of natural resources can be compared with a football game:
- In football, rules are set by a federation in order to determine mainly what cannot be done on the pitch;
- There are no official plans aimed at telling players what to do; this is the role of the coach as far as 'long-term' strategy is concerned. The players decide on actions on an ad hoc basis within this framework.

The comparison holds if one equates the federation to local governance, the coach to local leaders of user groups, and the players to different individuals or groups in the community.

**Table 4.6** *Some factors affecting decentralization for different sectors*

| Factors | Decentralization issues | | | Implications for agriculture |
| --- | --- | --- | --- | --- |
| | Agriculture | Water supply | Wildlife | |
| Economies of scale | Only for large processing/businesses | For large urban supplies | For large parks/major tourist ventures | Favours decentralization |
| Technical system | Small farmers – simple | Simple for basic systems – not major pump schemes | Complex knowledge of wildlife required | Favours decentralization |
| Complexity of environment | Complex – local knowledge needed for appropriate solutions | Simple | Wide range of issues – attitudes of community/tourists/demand for land | Favours decentralization |
| Diversity of clients, livelihood systems, etc | Depends on the area – often very diverse | Not a major issue – some tension between household, industrial and agricultural use | Diverse – community, employees, tourists, 'guardians of biodiversity' with multiple uses | Favours decentralization |
| Unit size/span of control | Dealing with many small farmers | Many villages/wells/pumps. Few large schemes | Whole area can be dealt with in conservation approach. Few/large parks | Varies – can mean efficacy of central mass media campaigns, but also need for support to local groups |
| Cross area boundaries | Large projects | Catchment management and major dams | Parks and conservation may well overlap | Only major projects. Favours decentralization |
| Stability | Depends on rate of change, but can change year to year | Depends on variability of rainfall | Depends on issues such as poaching, invasions, etc | Need for flexible systems to respond to dynamic situation. Favours decentralization |

| Factors | Decentralization issues | | | |
| --- | --- | --- | --- | --- |
| | *Agriculture* | *Water supply* | *Wildlife* | *Implications for agriculture* |
| *Political issues* | Land – eg land reform. Subsidies | Availability. Location of schemes. Charging structure | Highly political as users are often not local – if benefits do not accrue locally. Develop stakeholder forums | Issues such as land reform best handled centrally |
| *Fiscal issues* | Payment for extension? Subsidies on inputs/crops. Tax on land/crops. Free-rider issues | Payment for water – differential payments for types of users | How does revenue accrue locally? | Favours decentralization, although if tax is to be raised, local politicians may find it difficult. Maybe more possible with club goods |
| *Institutional issues* | Best handled locally – can be decentralized with specialized services provided centrally/regionally | Village water best locally, also municipal schemes. Bulk distribution perhaps regional. Catchment schemes regional | Must be local benefits and so link to local government and community structures. Parks: central/regional | Extension services can be handled locally; specialist services such as adaptive research better at regional level; basic research at national level |

---

**Box 4.15** *The Give a Dam Campaign, Zimbabwe*

This campaign started in late 1995, born out of the experiences of national and international NGOs involved with drought relief activities in Matabeleland South Province. NGOs, donors, Rural District Councils (RDCs) representing the communities in this province, and government agencies formed a consortium to construct small- to medium-sized dams to provide water for irrigation livestock and domestic use. The partners included:

| | | |
|---|---|---|
| Africare | Africa 2000 Network | Christian Care |
| CADEC | Dabane Trust | Matabeleland Development Foundation |
| Six RDCs | Government agencies | German Development Service |
| Oxfam USA | ORAP | Lutheran World Fund |
| Oxfam Canada | Evangelical Fellowship | UNDP and World Vision |

The campaign is jointly chaired by the Provincial Administrator of Matabeleland South and the UNDP Resident Representative.

**Achievements**

- 37 dams substantially completed;
- 8 dams under construction;
- 2 dams shelved for technical reasons;
- 2 dams breached due to cyclone-induced floods;
- 1 dam surveyed;
- 1 dam awaiting design;
- 3 dams awaiting funding;
- 1 site investigation under way;
- 17 irrigation schemes at various stages of construction (plus 6 under way).

**Training and ownership**

Each dam constructed under the campaign is managed by the dam management committee – selected by the communities. Its main responsibility is to help the community to participate fully in all dam-related activities. Training has been provided to 33 committees and to 47 facilitators drawn from the communities. Training is conducted at community, district, provincial and national levels and all training processes are facilitated by district training teams composed of extension workers from central and local government.

*Source:* PlanAfric (2001)

---

concerns. Here again, representation (as a step towards accountability) and cost-effectiveness (to avoid bureaucratic inflation) are key ingredients of success. In this respect, the eventual implementation of the new Traditional Leaders Act in Zimbabwe (Box 4.8) might provide useful insights.

## Coordinating different donors' initiatives

Lack of coordination between different donor-funded initiatives often exacerbates weak links between government line agencies. Frequently, efforts are made to establish coordinating committees, but practical difficulties (not least a lack of money to support regular meetings) means that such committees often fail to meet expectations.

### From integrated rural development to integral local development
Ribot (1999) argues that while the concept of integrated rural development *per se* had great merit, the mechanisms for integration were flawed – they were designed by outside agencies and often excluded local governments and local people from decision-making processes. He suggests that integrated rural development should be recast using a bottom-up approach – which he calls *integral local development* – with the aims of rehabilitating the role of local government in rural areas. Ribot argues that it is integral because it depends on authority that is integral to the community; environment is not separated from other matters of local government and all local funds are allocated according to local people's needs and aspirations, not according to the origin of the funds.

In this way, integrated local development is supported: if LGUs are entrusted with real powers and resources, and if they are made much more accountable to local communities. A similar focus on LGUs has been adopted by UNDP's LDF programme (see Box 4.13).

It is now accepted that outsiders should not drive the process of local development but in practice, donor-funded initiatives can help to test improved mechanisms for entrusting local governments with power and resources and for making them more accountable (eg see Box 4.16) through:

- (Regarding entrustment) embedding project structures in local government agencies, with an emphasis on capacity-building.
- (Regarding accountability) involving villagers in the monitoring – and control – of government staff activities associated with projects (eg notes on visits to villages, villagers being responsible for the delivery of fuel).
- Increasing the role of the concerned line agencies in project design and implementation can further enhance both entrustment and accountability.

### Capacity development

#### Mapping out capacity
Given the controversy over capacity issues (see 'Institutional realities', page 156), and the usually disappointing performance of capacity-building programmes, a step-by-step approach may be the best way forward:

1   Start by *agreeing on objectives* of development initiatives with local stakeholders.
2   *Clarify/(re)negotiate roles* between the main stakeholders.
3   The first two steps should serve as a basis for jointly *defining capacity needs*, as opposed to capacity-building objectives being decided mainly by donors and/or national governments.
4   Assess jointly with local stakeholders what can be achieved using the existing capacities, thus focusing on *capacity use* prior to capacity development.

---

**Box 4.16** *Project committees in Mali*

Under the GRN-GT Project in Mali, funded by GTZ, Local Development Committees (LDCs) comprise all line agencies operating in the area, chaired by the District Officer assisted by an expatriate adviser. Using PRA techniques, the LDCs develop technical plans. These are submitted to the project's national office which, in turn, is responsible for identifying the weaknesses of field operations and for developing a policy dialogue with both the LDCs and concerned government departments at central level.

Village committees have some control on the funds allocated by the project to line agencies, through the monitoring of LDCs. This type of mechanism encourages transparency in project management and more involvement of both LGUs and local communities. Problems include:

- committee members have different motives for participating;
- participation by line agencies in LDCs is often mainly linked to financial advantages;
- project managers are regularly confronted with requests for funds for *per diems*, training, study tours, etc, at the expense of other areas of management.

*Source:* Grosjean et al (1998)

---

5    Assess *where weaknesses lie* and what types of local partnerships might overcome them – encompassing *both technical and institutional capacities* so as to address the key *hidden institutional* problems related to good governance.
6    An initial build-up phase through a series of small-scale projects is often advisable.
7    A learning-by-doing philosophy should prevail in rural areas, at least in the early stages of capacity development.

## Training

Mayers and Kotey (1996), reviewing several community-based initiatives in Ghana, confirm that the provision of training is not necessarily a prerequisite for successful initiatives in local management. Assistance can be provided as needs arise and as part of a partnership process.

Training in managerial and planning matters is just as necessary as technical training. Linking training closely with the application of knowledge is a very effective way for trainees to gain tangible benefits from their newly acquired skills.

## Information: the right to know

The capacity of local institutions to manage natural resources depends upon the quality of the *information* they receive. Yet information on the formal roles of stakeholders (eg their '4Rs') as well as formal information about the resources to be managed is lacking or wholly inadequate (Dalal-Clayton and Dent, 2001).

In the wake of recent decentralization strategies, legislators and local communities have been increasingly concerned that communities should be informed about their new rights and duties – so that they might become more involved in the management of natural resources, albeit as stewards for the State.

In the case of the 4Rs, the interpretation of rules also plays a key role. Differences of interpretation often originate from differences in power – and threats to it. Therefore, information alone is insufficient to guarantee appropriate implementation of rules. It needs to be accompanied by the development of mechanisms which provide checks and balances.

## The role of the central government

Government policies on natural resources have a poor track record. Local communities have some comparative advantages in this respect (Baland and Platteau 1996):

- Local communities are *well informed about local ecological conditions*, although they may misapprehend the causes of some environmental changes.
- They are well *aware of local technical, economic and social conditions* as well as *cultural values*. Hence they are in a good position to devise management systems well adapted to their own needs, if not always to purely conservationist purposes.

---

**Box 4.17** *Possible role of central government in local forest management*

**Assistance and guidance** because it can more efficiently monitor the external effects of forest use. Main areas concerned include:

- perceiving environmental changes;
- convincing local groups that there are remedies to environmental stress;
- disseminating information on environmentally sound techniques, and sharing information both between the State and local groups and between groups.

**Provision of economic incentives** for conservation, especially where communities struggle to meet their basic needs and/or are at the mercy of powerful outside interests.

**Clarification of group territorial rights and provision of a legal framework** which enables user groups and their rights and benefits to be officially recognized.

**Protection** against broad-scale external pressures (eg pollution) and/or other economic sectors which central government is better able to respond to.

**Provision of formal rules for conflict-resolution** whenever locally derived rules are insufficient, especially in the case of conflicts between different communities and/or with broader-based stakeholders.

**Financial assistance** to complement the mobilization of local resources.

Provision of incentives and a framework to **link different decision-making levels**.

*Source:* Baland and Platteau (1996)

---

- Being relatively small, local communities *can easily adjust their local rules* of use to changing circumstances. But this depends on the nature and magnitude of the pressures placed on the resources and their livelihoods (eg externalities such as pollution are difficult to cope with).
- They usually have very *cost-effective mechanisms* to solve intra-community and interpersonal conflicts.
- *Self-monitoring* by the resource users themselves often proves cheaper and much more efficient than centralized control, as long as the community is convinced of its necessity.

Given these advantages of local control, the State would do well to relinquish power to local initiatives in resource management, even if they do not comply with any extant formal rules.

This does not mean that the State has no part to play in the management of natural resources. In addition to facilitating local initiatives in community management of natural resources, central government must play other important roles. Box 4.17 lists some of the roles of government that apply to forestry. For example, it should provide an overall vision for development that should be complemented by others at provincial and district levels.

# 5

# The way forward

In this concluding chapter, we summarize the lessons from half a century of professional natural resources surveys and development planning that are relevant today. The developing world (or *less developed countries* to use the term of the Brundtland report) is a very different place from what it was 50 years ago. Its human population has more than doubled from 1.7 billion to 4.5 billion and, at the same time, natural resources have been hugely degraded. In the next 50 years, its human population will increase to at least 6 billion (the lowest projection) and may well more than double again to 12.8 billion (the medium projection).

Fifty years ago, only 17 per cent of these people lived in towns and cities; now it is 35 per cent. It is impossible to forecast the extent of the shift to urban areas over the next 50 years. We can be sure that the rural population will still be increasing in absolute terms (Lutz, 1996) and the burgeoning cities will still depend entirely on their hinterlands for water supplies and waste disposal, even if successful cities are able to pay for food and fuel in global markets. The growing interdependence, with two-way flows of people, resources and activities between town and country means that rural planning in isolation from the wider economy is not realistic. There has never been a greater need for development planning and good management of natural resources, and this is no time for faint hearts.

This study demonstrates that there is a wealth of experience, and knowledge of the land, that points a way forward. The main problems that continue to dog development planning, and recommendations to overcome or at least mitigate them, are presented under the headings of 'Planning strategy', 'Principles of development planning', 'Natural resources surveys', 'Institutional support' and, finally, 'Implications for donors'.

## PLANNING STRATEGY

Development planning is unlikely to be effective in the longer term unless it operates within a truly domestically driven vision of the future – not imposed from outside, not tied to party, tribe or sect. To develop a vision that commands broad-based support is the most important national task for every country. All stakeholders need to participate and coordinated strategies are needed at national level, sub-national and local levels to work toward this vision so they pull together and are not at cross purposes. This has proved hard to achieve.

South Africa, Zimbabwe and Ghana demonstrate a spectrum of progress toward this goal. In South Africa, political accommodation is so recent that effective institutions of local government have yet to be built. Zimbabwe, a decade or more along this track, attempted to sweep away both the colonial imbalance of land holding and power and, also, long-established customary authority. But, after several false starts, there is still no common vision, no coordinated rural development strategy, and participation is taking place in a vacuum. In Ghana, the national vision (*Ghana Vision 2020*) and strategy were developed in 1994 and a start was made on implementation during 1995–2000. However, in 2001, a new government determined that the goals of *Ghana Vision 2020* could not be achieved in the planned time frame and proclaimed a new vision. Capacity at district level in Ghana is still wanting.

*Recommendation*

> *A vision of the future that commands wide support is a prerequisite of development and a necessary framework within which rural planning should operate. A process to develop this vision (eg round tables at local to national levels, future search conferences) must include all the stakeholders: resource users, other rural people including minorities, urban people, government, and NGOs. Sectoral, district, national and donor strategies should be coordinated to realise this vision.*

It is unrealistic and unhelpful to continue with rural planning and urban planning as separate entities. Urban centres are absolutely dependent on their hinterlands for water, fuel, food and waste disposal; there are important flows of people, goods and capital between them as well as more subtle social links; *urban* activities like industry occur in rural areas and *rural* activities like farming occur in urban areas. It certainly does not help that rural planning, land use planning and town planning are quite different schools that scarcely communicate with one another.

*Recommendation*

> *Urban–rural links should be recognized in the planning process and there should be dialogue between town planners, rural planners and economic planners.*

## PRINCIPLES OF DEVELOPMENT PLANNING

Planning has been very much a technocracy, in much the same way as natural resources surveying has been seen as a purely technical activity. Links between planning, natural resources surveys and policy have been and remain dysfunctional. They are different cultures and there is no common appreciation of their systems of values and modes of operation.

Macro-economic and commodity-based, business-style planning has colluded with a continued plunder of natural resources while failing to secure rural livelihoods. The concept of *sustainable livelihoods* provides a more holistic focus for development strategies. It concentrates attention on the need for secure livelihoods underpinned by sustainable management of natural resources and by human, social and financial capital – both capital generated within the rural area and that operating through external agencies. While the livelihoods perspective is described by fairly elaborate frameworks (Carney and Farrington, 1998; Bebbington, 1999; and Figure 1.1), it remains to translate the new perspective into practical decision-making. Planning surely holds the key to this conundrum – *if* it enables decision-makers at all levels to work effectively and equitably on the basis of sound information about the status of natural resources and the other forms of capital and the aspirations of the people.

The sustainable livelihoods concept is not a planning procedure in itself but is compatible with the *Steps in land use planning* (FAO, 1993; Dalal-Clayton and Dent, 2001) elaborated in Chapter 2 (Box 2.8).

Putting the security of rural livelihoods and the perspectives of rural people at the heart of the process will require fundamental shifts in what currently goes by the name of planning. It requires an open and inclusive planning process with open and equal access to information. This is not a neutral technical activity. It is highly political, as illustrated by the cases of Ghana (Botchie, 2000), South Africa (Khanya-mrc, 2000) and Zimbabwe (PlanAfric, 2000). It also takes time: ten years in the case of SARDEP in Namibia (Box 5.1).

*Recommendation*

> *Political as well as institutional reform will be required if rural people are to be allowed to genuinely plan for themselves. Given this opportunity, the concept of Sustainable rural livelihoods and the well-established Steps in land use planning offer a focus for policy and a procedure for planning.*

Strategic rural planning has now moved on beyond farm planning writ large. It has to integrate elements that have been the domain of quite separate professions – not just soil and water resource conservation, irrigation and drainage, but water resource allocation and total catchment management; development of sustainable production systems and their supporting infrastructure; and the development of the human capital of rural areas. Strategic rural planning seeks to achieve deliberate voluntary change of land use and management. This can only occur where there is knowledge, the capacity to change and the motivation to change:

- *Knowledge of the land* is essential at all levels of decision-making so that emerging problems and opportunities can be recognized. Good information is also needed about the social and economic consequences of change in

---

**Box 5.1** *Institutionalizing participation in Namibia: the case of SARDEP*

In late 1991, the Ministry of Agriculture, Water and Rural Development, with support from GTZ, launched the 10-year Sustainable Animal and Range Development Program (SARDEP) in the Communal Areas. The *aim* is to promote animal production and rangeland utilization that matches the natural resource base. All activities are based on a participatory approach and, in the orientation phase (lasting to late 1995), communal farmers have been assisted to plan, implement and monitor technical measures to improve livestock production and rangeland condition. Activities have included:

- Technical and organizational training for farmers as well as exposure to other situations, within and outside the country, to raise awareness about the current status of the natural resources.
- Encouraging community-based organizations (CBOs) cooperating with SARDEP to raise their own funds to reduce their dependence on governmental and external support.
- Assistance to CBOs to play a major role in coordinating the testing of technical measures at local level and promoting promising experiences among neighbouring communities.
- Introduction of *Participatory Rural Appraisal and Planning* to assist farmers in planning and to encourage self-help.
- Technical advice and financial assistance to farming communities to implement planned measures in order to qualify for grants; communities must make a cash contribution towards the envisaged costs.
- Assistance in finding additional support from governmental, non-governmental and private institutions.

A workshop with farmers and service delivery organizations (eg veterinarians, Departments of Water Affairs and Marketing) in 1994 showed that farmers have a clear vision of the future but have low self-help capacity. They don't know where to get support, while few of the service organizations can meet the farmers' demands. This led to SARDEP adopting a strategy to:

- support CBOs;
- identify problems and create a policy framework conducive to sustainable development:
  - people from communal areas and policy-makers have been brought together in workshops to discuss issues – the aim is to 'take policy-makers to the land',
  - studies have been provided to policy-makers to inform their thinking,
  - SARDEP has played a constructive role in developing the National Agricultural Policy and the Drought Strategy and Policy, and has provided information to the Ministry of Lands during the preparation of the Communal Lands Bill;
- support organizations in re-orienting services towards the needs of communal farmers.

*Sources:* Sullivan (2001) citing Laue and Kruger (1995); Bertis Kruger (pers. comm.)

---

land use and management, and of not changing. There is no single way to bring about this exchange of knowledge between all stakeholders, and a range of approaches will need to be tried, modified and repeated.

---

**Box 5.2** *Principles of land use planning: recommendations*

1   Planning should not be undertaken exclusively by professionals, remote from the area concerned. If it is to be successful, a plan needs to be developed in partnership with all of those with a legitimate interest, particularly residents of the area and those whose livelihoods depend on its resources. Identify these people first, and establish a mechanism for them to participate in the planning process.
2   Acknowledge the existence of conflicting interests in developing, implementing and benefiting from land use plans. Develop processes to deal with this. The needs and goals of all the interest groups should be clarified in the light of the aims of the plan.
3   Address social issues, especially land tenure and access to resources, as well as physical or environmental issues.
4   To the extent possible, try to reach consensus, taking particular care to include marginalized groups, eg women and minorities.
5   Consensus-building and meaningful negotiation require equal access to information about the issues, problems and development options. Build on:
    −   indigenous systems of local knowledge, land use and planning, taking care to retain their diversity and flexibility;
    −   the experience and expertise of other sectors and NGOs.
6   Common property (or un-priced) resources such as land, water, pastures, forest and wildlife have important economic values and are not infinitely substitutable. There needs to be an accounting system to assess depreciation of these natural resources and a mechanism to ensure their sustainable management, otherwise they are likely to be exploited to the point where the system is destroyed.
7   Build and support local institutions that can manage common property resources and devolve authority to them.

---

- *Capacity to change.* Lack of time, people, management skills, appropriate institutional structures (tenure, laws or decision-making systems), equipment and money are all constraints. Financial and managerial resources are needed for activities that do not give a quick return and, often, benefit urban people more than those in rural areas, eg safeguarding water resources.
- *Motivation to change.* Education, information and persuasion will be effective only if the change interests and benefits the people for whom change is deemed desirable, and where such change is socially acceptable.

Our recommendations are summarized in seven principles in Box 5.2.

## NATURAL RESOURCES SURVEYS

It is impossible to overstate the need for reliable information about natural resources, both their present condition and how this will respond to the various management options. More than five thousand years of recorded history, quite apart from our own half century's professional experience, attest to the price of ignorance.

The information needed is not merely inventory but requires a sound knowledge of processes, so that the effects of management can be predicted and sustainable systems of management devised, and this is not come by lightly. We have to ask why the lessons of history are still not learnt, by policy-makers in particular; and why basic information about natural resources, gathered by professional teams in the 1950s to 1980s, is being fecklessly discarded and the cadre of natural resources professionals is being allowed to atrophy.

Certainly, the standard information provided by natural resources surveys doesn't fit into the decision-makers' economic concepts and can't be used directly to assess opportunities or risk. It is asking a lot of policy-makers, planners and land users to bridge the gap between a soil map, biodiversity inventory or table of climatic data, and an assessment of the various development options and forecasts of their outcomes. Where interpretations of basic data have been provided, they have been taken off-the-peg, not tailored to the decision-makers' requirements. This latter point can be applied equally to a great deal of other specialist information (social, demographic, economic, financial, etc).

Taking specific criticisms of natural resources surveys in sequence:

1 'What I'm wanting I'm not getting! What I'm getting I'm not wanting.'

*Recommendations*

> *Natural resources agencies must establish who their customers are. Then there must be dialogue with them to establish the precise information needed.*

> *Blunderbuss surveys should not be continued. If the immediate use of data is not known, don't collect them.*

> *Natural resources specialists have to go the extra mile to tailor their outputs to fit into the way decisions are taken. They must recognize the format in which the information is needed and the matrix of other information into which their own contribution must be dovetailed.*

> *Simulation models can provide production and input data for economic models, and forecasts in terms of probability. Existing survey data are rarely adequate for simulation models but expert knowledge can be used to devise 'transfer functions' that relate existing data to the parameters required.*

2 Almost universally, we face information gaps. Most often, there is a gap between the scale at which information is needed and the scale at which it can be provided quickly and cheaply. Survey coverage and recording networks are incomplete at any level of detail and there is no strategy for completion, no continuity of survey effort and no continuity of staff. As a result, methods are not compatible, expertise is not built up, and institutional memory is short.

*Recommendation*

> *Greater investment in land resource information would be beneficial but, at the same time, we need procedures for survey and for maintenance of the database that can be implemented by imperfect and modestly funded institutions.*

3    Data are of variable quality and detail. This limits the use that can be made of the better data. For example, soil descriptions gathered by land systems surveys are often cursory (eg 'reddish brown earth') so little interpretation is possible. Other data are generated by remote sensing without adequate (or any) field checks. Yet recent developments in airborne and satellite-borne sensors and positioning systems offer unprecedented opportunities for very cost-effective survey of broad tracts of land, at the same time offering levels of detail that can scarcely be matched by ground survey. It requires only rigorous ground control and the professional expertise built up by this field calibration to take advantage of the new systems.

*Recommendation*

> *Surveyors must develop and stick to rigorous quality control. Users should be able to ascertain the date and accuracy of data from the files. GIS managers must ascertain the compatibility of different data sets. Those commissioning surveys should not neglect the need for fieldwork, nor underestimate the time and cost involved.*

4    It is difficult to disentangle primary, factual data from interpretations.

*Recommendation*

> *Surveys should make explicit the distinction between 'What is the land like?' and 'What is it good for?'. Data banks should store these different kinds of data in separate files, eg 'physical' files for climatic data, soil data etc; 'use' files for the results of experimentation and practice, eg crop yields, water yields; and 'interpretation' files for judgements, eg land suitability.*

5    There is a critical shortage of competent land resources specialists, both in specialist institutions and in line ministries and local government able to provide a service to policy-makers, planners and land users. The lack of a career structure in land use planning is a major constraint. The secondment of staff from established institutions is universally unsatisfactory. While information gaps can be plugged by overseas-funded and -managed survey programmes, these do not maintain the database for iterative use or provide new interpretations at a later date.

*Recommendation*

> *An attractive career structure is necessary to attract and retain able people. Technical training in natural resources survey and land use planning is urgently needed for professional staff in developing countries but it will be more realistic in the physical, social and economic environment of the home country or a regional centre: training in rich countries invariably emphasizes hi-tech and laboratory-based techniques that are inappropriate where state-of-the-art technology cannot be afforded.*

6   Indigenous, local knowledge of the land has long been ignored by technical specialists and policy-makers alike. Following Robert Chambers' *Farmer First* ideas, a tidal wave of people-oriented approaches has swept the field to the extent that hard-won, significant physical data are ignored, even dismissed, in favour of 'participation' and 'bottom-up planning'. But no amount of participation can make it rain.

*Recommendation*

> *Links should be forged between top-down and bottom-up planning. Professional training should encompass BOTH the techniques of participatory inquiry and those of the natural sciences, so that information can be gleaned, as appropriate, from both approaches and combined.*

7   Natural resources information is incomprehensible: confounded by jargon and fogged by the sheer mass of data. The comprehension barrier is raised as soon as we break the links with logic that is based on common sense and everyday experience.

*Recommendation*

> *Apart from using plain words and few, it is necessary to communicate knowledge of the land by analogy with common experience, even if this takes some time and trouble.*

> *A lot of jargon and complexity arose from the needs of manual systems of storage and retrieval of information through maps and abstract hierarchical classifications. This need has been swept away by computer-based information technology, although the new Nirvana has gatekeepers of its own.*

8   'We have been pouring information into the sand!'

*Recommendation*

> *Specialists must carry their information to the point of decision, otherwise much is lost or corrupted along the way. Synthesis and generalization from*

*factual detail cannot be accomplished by generalist planners and policy-makers, only by a master of the information.*

In spite of all the problems listed above, baseline surveys of natural resources remain indispensable to sustainable management of natural resources. They are needed to assess the present and they are critical for future monitoring. But they are usually divorced from policy-making at any level and from stakeholder and community participation.

All the perceived shortcomings can be rectified so that baseline natural resources information can be integrated in a timely fashion with other aspects of rural planning, drawing upon both specialist technical knowledge and local knowledge. Compared with the cost of ignorance and the cost of even the most basic infrastructure development, the price of this knowledge is modest indeed.

## INSTITUTIONAL SUPPORT

The State is not a good manager of natural resources. All the evidence points to the need for active involvement of local people in managing the resources they depend upon, and in planning their own development. But they need institutional support from various levels of government:

1  Platforms for decision-making at local and district level where stakeholders can meet on equal terms to negotiate development goals and the allocation of resources.
2  Responsive, effective services provided at district level.
3  In big countries, coordination across districts and provision of specialist services at provincial level. In very small countries, cooperation within the larger region.
4  Strategic direction and redistribution of resources at national level.

This is a subject of its own, with a huge literature that we have not been able to deal with completely. Our recommendations are therefore about what needs to be done rather than how it should be done.

### Local (community) level

At this level, there is rarely any institution with a mandate for development planning and, consequently, little or no capacity or experience. However, there are many examples of good management of resources such as irrigation water, drainage, pastures, forest and wildlife by local communities. It should be possible to build on these existing management structures.

*Recommendations*

   1   *Where institutions have to be built from scratch, critical elements will be:*

> — *management of the local area by the local community, focusing on what people can do for themselves – not just bidding for handouts;*
> — *local structures developing local plans that contribute, in turn, to a more comprehensive district plan.*

2   *Features of successful management include:*
> — *participation by resource users and other major stakeholders (eg partnership institutions between user and government agencies);*
> — *transparency in decision-making;*
> — *financial autonomy, eg funding through user fees;*
> — *adequate monitoring to identify rule-breakers and assess changes in the status of resources;*
> — *timely, effective conflict-resolution mechanisms;*
> — *sufficient knowledge and skills;*
> — *matching of institutions, responsibilities and scales of operation.*

## Local government level (eg district)

At this level, the key functions are:

- Delivery of local services (schools, clinics, technical advice, etc).
- Strategic planning of infrastructure and services. Allocation of rights to water, common grazing, timber, wildlife and other resources, according to national legislation – and monitoring and policing of this use.
- Coordination of local development plans.

Decentralization in some form is promoted as a cure for all the ills of rural development. It promises:

- participation and empowerment of rural people;
- more equitable distribution of benefits;
- improved local accountability;
- more effective delivery of services.

It has usually failed to deliver – because central governments have been reluctant to cede any real power, resources have not been commensurate with responsibilities, and because local governments have succumbed to *invisible institutions* (patronage, rent-seeking, antipathy to participation, etc).

To build adequate technical and institutional capacity at the district level is a formidable task. There are weaknesses in technical capacity to do the planning and inadequate financial resources to implement and monitor the plans.

Nearly all experience worldwide shows that devolution of service delivery to the district level improves effectiveness and responsiveness to local needs. However, this requires devolution of commensurate powers and resources to the districts, accountable financial autonomy and a big investment in professional capacity at the district level. This last point applies equally to professional capacity for development planning. At the moment, districts do not have this capacity and cannot build it unless their staff are accountable at the district level, not to a higher level.

Capacity is also weak in representation, negotiation, transparent procedures and mechanisms. Support is needed in each of these fields to enable rural people to appropriate the process and use it to meet their own development priorities.

*Recommendations*

> *If they are to be effective, decentralized systems must have:*
> * *devolution of enough power to exert significant influence over affairs;*
> * *reliable accountability mechanisms;*
> * *finance sufficient to accomplish important tasks;*
> * *adequate technical and institutional capacity to accomplish those tasks.*
> *Outside assistance should take care not to hijack the process of capacity-building by imposing an external perspective.*

> *Capacity development might proceed step by step, eg:*
> 1  *agreement between stakeholders about the development goals;*
> 2  *clarification and negotiation of roles between the stakeholders;*
> 3  *definition of capacity needs by the stakeholders, as opposed to capacity-building objectives decided by donors or national government;*
> 4  *assessment of what can be achieved using existing capacities;*
> 5  *identification of weaknesses and what local partnerships might overcome them.*

## Intermediate level (eg province, region)

In some countries, the switch of responsibilities to districts has much reduced the role of the province. Sometimes this has created a support gap. Certainly in big or very diverse countries, there are several continuing functions for the province;

* Capacity-building for districts, because there are able staff at the province level.
* Technical support where there is not enough demand or supply for a full professional complement at district level, eg for an irrigation engineer.
* Audit of local government with sanctions against poor performance.
* Coordination of development plans across districts, and development of strategic opportunities at province level (eg for tourism, water/power resources).

## National level

The push toward decentralization is often seen as a threat to the power of central government and the budgets of line ministries. In an important sense this is true, but it is still in the long-term interest of government staff and politicians to improve service delivery – they will get the credit for this.

Decentralization is not so much a loss of role for the centre but a change of role, for example:

- An increased demand for higher-level, specialist services which will be provided by line ministries, possibly at the province level.
- Increased information flow, replacing control by signature (or many signatures) by control through monitoring and by ensuring financial and legal discipline.
- Strengthened policy-making, taking a more strategic approach to development, seeking points where State initiative and support can make the most impact.

Situations in rural areas are very diverse and often complex, calling for creativity and flexible approaches. On the face of it, it should be helpful to develop interim management measures, and test them over time by trial and error. A range of possible solutions should always be assessed. As a result of this participatory learning approach, institutions should adapt to realities rather than the reverse. Such an approach requires a shift to a more locally based strategy to regulate the use of natural resources.

All experience shows that the more effective government approaches are:

- *Progressive*, that is tenure practices, regulations and institutions adapted to stakeholders' needs and wishes.
- *Selective*, acknowledging the need for a transition phase.
- *Pragmatic*, supporting and improving upon rather than replacing, existing practices.

*Recommendations*

> *Governments might consider a contractual approach between themselves and local communities, whereby mutual obligations concerning the use of natural resources are negotiated (Karsenty, 1996). This would lead to a gradual shift from:*
>
> > *laws/institutions* → *tools* → *project* → *participation;*
> >
> > *towards:*
> >
> > *general legal principles* → *negotiation* → *definition of long-term objectives* → *common choice of instruments and setting up of local institutions* → *evolution and adaptation.*
>
> *Development should be based on a process of continuous and mutual learning, allowing for:*
>
> > - *experimentation with a variety of platforms for decision-making at local, district and higher levels;*
> > - *confidence-building between stakeholders;*

- *better communication, through information, negotiation and mediation; and, finally,*
- *time for the process to materialize into effective policies.*

## CONCLUSIONS

Plans made for the rural areas of developing countries are often not implemented and sectoral projects exist in a policy vacuum. Commonly, this state of affairs stems from the mandates of the responsible authorities that lack either enough authority, or enough resources, or both.

Any number of case studies show that rural planning *can be* a way for local stakeholders to establish their priorities, bring them into the public domain, and negotiate plans and action to realize these priorities. This acknowledges the political dimension of local planning. But if rural planning is to make a difference to the well-being of rural peoples and contribute to more sustainable management of natural resources, each of the following issues must be tackled.

Local strategic planning requires reliable information about the condition and trends of natural resources, and about social and economic conditions. Methods to gather, synthesize and interpret this information are well established. Methods and mechanisms also exist to enable the participation of stakeholders in the planning process. It should be obvious that there is also a requirement for:

- skilled and dedicated people to use these methods;
- a planning framework within which they can be brought together; and for
- financial resources sufficient to do the job at the local level.

At the local government level (eg district), which is the key level for both service delivery and local strategic planning, knowledge, professional and financial resources are not sufficient for the task. We are dealing with poor countries where these resources are meagre in any case but, in most cases, the district level has not been entrusted with responsibility for strategic planning or the raising of money (eg through taxation).

Nevertheless, we conclude that it is at the district level that the most immediate needs of both rural and urban communities can best be met. This subsidiarity does not entail so much a loss of role by central government and its institutions as a shift to a more strategic role.

### Implications for donors

The implication for donors is clear. Assistance for rural planning in a particular country will require a measured, structured response. The first step should be an assessment of current rural planning arrangements (planning framework, institutional roles and responsibilities, skills-base, etc) and an assessment of needs. Nationals should take the lead in such an assessment, as part of the raising of awareness of the issues and possible responses. This study indicates that donor assistance must encompass considerable investment in training, building capacity and skills, and providing incentives for established bureaucracies to change. All this will need a long-term commitment. Without these developments, support for projects which arise from flawed planning processes will not, except by luck, lead to improvements in livelihoods.

A key issue for donors is to identify the possible *entry points* to assist capacity-building. Possible initial activities should be developed locally from a consultative *needs assessment*. This provides all primary stakeholders with the opportunity to express their aspirations and concerns and should be the first stage of any planning process. Common sense suggests that the initial activities should be those which have a reasonable chance of success. Success breeds self-confidence and increased self-confidence, in turn, enables people to tackle more difficult issues.

In this way, a *process* that builds local capacity is also initiated. The planning process does much more than merely produce a plan. More importantly, it also builds people's skills and ability to plan and follow up planning with concerted actions. Initial activities should be selected that stretch existing capacity, but not by so much that it breaks down.

The classic development assistance or NGO 'project' may not be the only possible entry point. Local, community-based initiatives may also be identified during a needs assessment or scoping study and these may provide entry to a wider planning process.

There is always a pressure to provide a *blueprint* to structure planning approaches. We do not believe that blueprints are useful to rural planning. They tend to lead to routine or mechanical adherence to procedures, leaving little room to respond to the realities on the ground, or to think creatively and innovate.

Current thinking on ways of engaging in the development and implementation of national strategies for sustainable development (as discussed in Chapter 2 'Some planning responses to the challenge of sustainable development', page 68) offers a framework for planning which is equally valid at more local levels – one which is centred on participatory processes of analysis, debate, capacity-strengthening, planning and action. Following this thinking, Tanzib Chowdhury (pers. comm.) suggests that it might be possible to build upon the new partnership approach adopted by the World Bank's initiative on city development strategies (CDSs) and promote district development strategies (DDSs) (see Box 5.3).

This is not an entirely new idea. Indeed, aspirations to such a dynamic rural planning system can be seen in many developing countries. For example, the decentralized planning system that was growing out of Ghana's *Vision 2020*, the integrated development plans of South Africa and the rural master plans in Zimbabwe all aspire to be tools for coordination, management and gathering community inputs. DDSs could strengthen such emerging planning systems that are now faltering because of institutional problems and lack of technical capacity.

While there is a lot of rhetoric promoting participatory, integrated and decentralized planning systems, practice has fallen short. Sectoral planning

**Box 5.3** *From city to district development strategies*

The World Bank has launched a new *partnership approach* to city assistance – the city development strategy (CDS). Recent trends in urban affairs (decentralization, democratization, emphasis on participation in governance), the limited impact of individual urban projects and sectoral loans, and the recognition that cities are contributing increasingly to national economies have led to the emergence of the CDS concept. The CDS, which is being piloted in Bangladesh, China, Columbia, the Phillipines, South Africa, Sri Lanka, Thailand, Uganda and Vietnam, aims to analyse the key urban growth issues (economic, environmental, poverty, governance) seen from the perspective of the city stakeholders, to consult and advise on priorities, and to suggest priority assistance and a future work programme. The key elements include: extensive participatory activities with large numbers of stakeholders; in-depth examination of the composition of the city economy; and the development of a coordinated framework for donor assistance. The CDS would not necessarily be equally comprehensive for every city as the impetus for the strategy exercise can arise from various priorities perceived by the stakeholders.

Rural areas are subject to the same forces of decentralization and democratization, yet district authorities don't have the capacity to plan effectively and engage communities in the process. Projects and departments work on a sectoral basis. Usually, there is a very limited understanding of how the local rural economy works and its relationship with the urban economy. A model similar to CDS could be developed in rural areas. Put simply, a district development strategy (DDS) would be a framework for doing business at the local level in a more integrated, coordinated, participatory way that pays particular attention to promoting sustainable livelihoods.

The emerging planning systems in Ghana, South Africa and Zimbabwe are all trying to move in this direction.

Some donor assistance is also being directed to promote and support this process (eg in South Africa, GTZ is supporting an Integrated Planning Support package). But there are considerable gaps and obstacles, as this overview of international experience clearly demonstrates. A DDS would be concerned with defining this approach more sharply, concentrating efforts in a few localities to develop good practice, and targeting assistance in a significantly more coordinated way.

The DDS would seek broad coalitions of local stakeholders and development partners, both local and international. Together, these would develop a strategy for a particular rural district that reflects a common understanding of the constraints and prospects (an analytical assessment) and a shared vision of goals, priorities and requirements (a strategic plan of action). The DDS would be both a process and a product identifying ways of moving toward sustainable development.

The process would be defined by the district but might involve the following stages:

- *Preparatory phase:* A quick assessment of the readiness of the district to engage in the process, and the key concerns of district officials and other key local stakeholders
- *Analytical phase:* An in-depth analysis of the structure and trends in the local economy, the various obstacles – institutional, financial, environmental and social – which may impede progress with a particular emphasis on inter-relationships among issues and stakeholders. A number of options would be developed according to the analysis
- *Consensus-building phase:* Setting priorities, building consensus, making decisions and identifying sources of assistance. This would include determining how local and international partners can help the district achieve its goals.

> The process of developing and implementing the strategy should be directed and owned by the district. Donors should facilitate the process and, perhaps, provide some technical inputs if requested. A DDS would be expected to define development priorities for the district but could go further and identify potential sources of assistance to realize strategic components. This process could help to coordinate donor assistance in a more effective manner.

systems continue to work against efforts to promote an integrated approach. Donors continue to respond to rural development needs with individual projects or sector loans that do not address the multi-sectoral, multi-dimensional problems faced by rural areas (including rural–urban interactions). The lack of an enabling framework (political, administrative and fiscal decentralization) and the lack of an empowering framework for community participation remain major constraints. Development plans are still largely descriptive; factual rather than analytical; and do not explore the inter-relationships between different sectors, issues and stakeholders.

## Uncertainties

Development planning is dogged by uncertainties:

- Are bureaucrats willing to do things differently? – Will they think and behave in new, open, participatory ways that provide for dialogue and consensus-building to agree goals and how to get there? There is a need to identify those motivations that will encourage bureaucrats to work in new ways.
- Are institutions capable of working in support of each other to achieve cross-sectoral integration and synchronization? There is a need to identify and support, in each situation, the constructive institutional relationships that exist; and ways must be identified to face up to the initial severe capacity constraints and to build the required capacity.

Perhaps most critically:

- Is there the political will to drive the needed changes?

Finally:

- What is the appropriate role for government? This needs to be clarified, particularly in relation to private enterprise which is an important resource (in many countries virtually the only resource, albeit it at a low level) to power development.

# References

Adnan, S, A Barrett, S M Nurul Alam and A Brustinov (1992) *People's Participation: NGOs and the Flood Action Plan*, Research and Advisory Services, Dhaka

Agarwal, A and S Narain (1990) *Village Ecosystem Planning*, Dryland Issue Paper No 16, International Institute for Environment and Development, London

Agarwal A and S Narain (1992) 'Community environmental governance: the village republic'. Paper presented to the SAREC Workshop on People's Participation in the Management of Natural Resources, Stockholm

Aguirre Beltrán, G (1962) *Regiones de Refugio: El Desarollo de la Comunidad y el Proceso Dominical en Mestizoémrica*, Instituto Nacional Indigenista, Mexico, DF

Aitken, J F (1983) 'Relationships between yield of sugar cane and soil mapping units, and the implications for land classification'. *Soil Survey and Land Evaluation* 3(1):1–9

Anyang' Nyong'o, P (1997) 'Institutionalisation of democratic governance in sub-Saharan Africa'. ECDPM Working Paper No 36, European Centre for Development Policy Management, Maastricht, June

Arnold, J E M and P A Dewees (eds) (1995) *Tree Management in Farmer Strategies: Responses to Agricultural Intensification*, Oxford University Press, Oxford

Atkinson, D and M Inge (1997) *Rural Local Government in the Free State*, Land and Agriculture Policy Centre, Johannesburg

Austin, M P and K D Cocks (1978) *Land Use on the South Coast of New South Wales: A Study in the Methods of Acquiring and Using Information to Analyse Regional Land Use Options*, 4 vols, CSIRO, Melbourne

Baburoglu, O N and M A Garr (1992) 'Search conference methodology for practitioners'. In: M R Weisford (ed) *Discovering Common Ground*, Berrett-Koehler, San Francisco, pp77–81

Backhaus, C and R Wagachchi (1995) *Only Playing with Beans? Participatory Approaches in Large-Scale Government Programmes*. PLA Notes No 24, Sustainable Agriculture Programme, International Institute for Environment and Development, London

Baker, J (1995) 'Survival and accumulation strategies at the rural-urban interface in north-west Tanzania'. *Environment and Urbanization* 7(1):117–132

Baland, J M and J P Platteau (1996) *Halting Degradation of Natural Resources: Is There a Role of Rural Communities?* Clarendon Press, Oxford

Banerjee, A and E Lutz (1996) 'Analysis of World Bank and GEF projects'. In: E Lutz and J Caldecott (eds) *Decentralisation and Biodiversity Conservation*, a World Bank Symposium, pp145–54

Bass, S and R Hearne (1997) *Private Sector Forestry: A Review of Instruments for Ensuring Sustainability*, Forestry and Land Use Series No 11, International Institute for Environment and Development, London

Bass, S and P Shah (1994) 'Participation in sustainable development strategies; with a case study of joint forest management in India'. Presentation to IUCN General Assembly, Workshop, 9 January, Buenos Aires

Bass, S, D B Dalal-Clayton and J Pretty (1995) *Participation in Strategies for Sustainable Development*, Environmental Planning Issues No 7, International Institute for Environment and Development, London

Bass, S, W Hawthorne and C Hughes (1998) 'Forests, biodiversity and livelihoods: linking policy and practice'. An Issues Paper for DFID, December (second draft)

Bawden, M G and D M Carroll (1968) *The Land Resources of Lesotho*, Land Resource Study No 7, Land Resources Development Centre, Ministry of Overseas Development, Surbiton

Bawden, M G, D M Carroll and P Tuley (1972) *The Land Resources of North East Nigeria, Volume 3: The Land Systems*, Land Resource Study No 9, Land Resources Development Centre, Ministry of Overseas Development, Surbiton

Bebbington, A (1999) *Capitals and Capabilities: A Framework for Analysing Peasant Viability, Rural Livelihoods and Poverty in the Andes*, International Institute for Environment and Development, London

Bebbington, A, A Kopp and D Rubinoff (1997) 'From chaos to strength? Social capital, rural peoples organisations and sustainable rural development'. Paper prepared for the FAO Workshop on Pluralism, Management and Rural Development, 9–12 December

Beek, K J, P A Burrough and D E McCormack (eds) (1987) *Quantified Land Evaluation*, ITC Publication 6, Enschede

Blair, H (1989) *Can Rural Development be Financed From Below? Local Resource Mobilization in Bangladesh*, Dhaka University Press Ltd

Blair, H (1997) 'Spreading power to the periphery: a USAID assessment of democratic local governance'. Paper prepared for the Technical Consultation on Decentralisation, Food and Agriculture Organization of the United Nations, Rome, 16–18 December

Bliss, F (1999) 'Theory and practice of participatory project planning; good concepts but difficult to realise'. *Development and Change* **1**(99):20–3

Bond, R (1998) *Lessons from the Large Scale Application of Process Approaches in Sri Lanka*, Gatekeeper Series No 75, International Institute for Environment and Development, London

Bonnet, B (1995) 'Instances décentralisées de décision, de régulation et de contrôle – Eléments de réflexion sur quelques expériences en cours dans le cadre des projets de gestion de terroir ou de développement local'. In: *Rapport de la Journée d'Etude 1995: Le Développement Local*, Institut de Recherches et d'Applications des Méthodes de Developpement, Montpelier

Botchie, G (2000) *Rural District Planning in Ghana: A Case Study*, Environmental Planning Issues No 21, International Institute for Environment and Development, London

Brammer, H (1983) *Manual of Upazilla Land Use Planning*, Min. Local Govt. Rural Development and Cooperation, Dacca

Bryceson, D F (1997) 'De-agrarianisation in sub-Saharan Africa: acknowledging the inevitable'. In: D F Bryceson and V Jamal (eds) *Farewell to Farms: De-Agrarianisation and Employment in Africa*, Africa Studies Centre Leiden, Research Series 1997/10, Ashgate, Aldershot, pp3–20

Bunch, R (1983) *Two Ears of Corn: A Guide to People-Centred Agricultural Improvement*, World Neighbors, Oklahoma City

Burnham, C P, A C Shinn and V J Varcoe (1987) 'Crop yields in relation to classes of soil and agricultural land classification grade in S.E. England'. *Soil Survey and Land Evaluation* **7**:95–100

Caldecott, J (1996) 'Good governance in model and real countries'. In: E Lutz and J Caldecott (eds) *Decentralisation and Biodiversity Conservation*, a World Bank Symposium, pp139–44

Campbell, A (1994) *Community First Landcare in Australia*, Gatekeeper Series No 42, Sustainable Agriculture Programme, International Institute for Environment and Development, London

Campbell, B, N Byron, P Hobane, F Matose, E Madzudzo and L Wily (1996) 'Taking CAMPFIRE beyond wildlife: what is the potential?' Paper presented at the Pan African

Symposium on Sustainable Use of Natural Resources and Community Participation, Harare, Zimbabwe, 24–27 June

Carew-Reid, J, R Prescott-Allen, S Bass and D B Dalal-Clayton (1994) *Strategies for National Sustainable Development: A Handbook for their Planning and Implementation*, International Institute for Environment and Development, London, and World Conservation Union, Gland, in association with Earthscan Publications, London

Carley, M (1994) *Policy Management Systems and Methods of Analysis for Sustainable Agriculture and Rural Development*, International Institute for Environment and Development, London, and UN Food and Agriculture Organisation, Rome

Carney, D (ed) (1998) 'Sustainable rural livelihoods: what contribution can we make?'. Paper presented at the Department for International Development's Natural Resource Advisers' Conference, July, Department for International Development, London

Carney, D and J Farrington (1998) *Natural Resource Management and Institutional Change*, Routledge Research/ODI Development Policy Studies, Routledge

Chambers, R (1992) 'Methods for Analysis by Farmers: The Professional Challenge'. Paper for 12th Annual Symposium of Association for Farming Systems Research and Extension, Michigan State University, 13–18 September

Chambers, R (1995) *Making the Best of Going to Scale*, PRA Notes No 24, Sustainable Agriculture Programme, International Institute for Environment and Development, London

Chambers, R (1997) *Whose Reality Counts? Putting the last first*, IT Publications, London

Chang, L and P A Burrough (1987) 'Fuzzy reasoning: a new quantitative aid for land evaluation'. *Soil Survey and Land Evaluation* 7:69–80

Chenery, H, M S Ahiuwalia, C G Bell, et al (1974) *Redistribution with Growth*, Oxford University Press (for the World Bank)

Clarke, J E (1983) *Master Plans for National Parks and Wildlife Management*, 4 vols. Dept. National Parks and Wildlife, Lilongwe, Malawi

CMDT (1991) *Proposition pour une méthode d'analyse villageoise: Projet gestion de terroir, San-Koutiala*, Compagnie Malienne pour le Développement des Textiles, Mali

Conway, G R, J A McCracken and J N Pretty (1987) *Training Notes for Agroecosystem Analysis and Rapid Rural Appraisal*, International Institute for Environment and Development, London

Club du Sahel/OECD (1998) *Decentralisation and local capacity building in West Africa: results of the PADLOS-Education Study*, August

Colfer, C J P (1995) 'Who counts most in sustainable forest management?' CIFOR Working Paper No 7, October, Center for International Forestry Research, Bogor, Indonesia

Coode and Partners (1979) *The Gambia Estuary Barrage Study – Stage II*, Coode and Partners/Govt. of The Gambia/Min. Overseas Development, London

Corker, I R (1982) *Human Carrying Capacity Assessment Model for Tabora Region*, Land Resources Development Centre, Project Record 65, Surbiton

Corker, I R (1983) *Land Use Planning Handbook. Tabora Rural Integrated Development Programme: Land Use Component*, Project Record No 65, Land Resources Development Centre, London

Cornes, R and T Sandler (1986) *The Theory of Externalities, Public Goods and Club Goods*, Cambridge University Press, Cambridge

Craswell, E, M Rais and J Dumanski (1996) 'Resource management domains as a vehicle for sustainable development'. IBSRAM Working Paper, IBSRAM, Bangkok

Curtis, A and T de Lacy (1997) 'Examining the assumptions underlying landcare'. In: S Lockie and F Vanclay (eds) *Critical Landcare*, Centre for Rural Social Research, Charles Sturt University, Wagga Wagga, NSW, Australia, pp185–200

Dalal-Clayton, D B (1993) *Modified EIA and Indicators of Sustainability: First Steps Towards Sustainability Analysis*, Environmental Planning Issues No 1, International Institute for Environment and Development, London

Dalal-Clayton, D B and B Child (2002) *Lessons from Luangwa: A Historical Review of the Luangwa Integrated Resource Development Project, Zambia,* LIRDP and International Institute for Environment and Development, London

Dalal-Clayton, D B and D L Dent (1993) *Surveys, Plans and People: A Review of Land Resource Information and its Use in Developing Countries,* Environmental Planning Issues No 2, International Institute for Environment and Development, London

Dalal-Clayton, D B and D L Dent (2001) *Knowledge of the Land: Land Resources Information and its Use in Rural Development* Oxford University Press, Oxford

Dalal-Clayton, D B and B Sadler (1998a) *The Application of Strategic Environmental Assessment in Developing Countries: Recent Experience and Future Prospects, Including its Role in Sustainable Development Strategies,* Environmental Planning Issues No 18, International Institute for Environment and Development, London

Dalal-Clayton, D B and B Sadler (1998b) 'Strategic environmental assessment: a rapidly evolving approach'. In: A Donnelly, D B Dalal-Clayton and R Hughes (eds) *A Directory of Impact Assessment Guidelines. Second Edition,* International Institute for Environment and Development, London

Dalal-Clayton, D B, S Bass, N Robins and K Swiderska (1998) *Rethinking Sustainable Development Strategies: Promoting Strategic Analysis, Debate and Action,* Environmental Planning Issues No 19, International Institute for Environment and Development, London

Dalal-Clayton, D B, D Dent and O Dubois (2000) *Rural Planning in the Developing World with a Special Focus on Natural Resources: Lessons Learned and Potential Contributions to Sustainable Livelihoods,* Environmental Planning Issues No 20, International Institute for Environment and Development, London

Dalal-Clayton, D B, K Swiderska and S Bass (eds) (2002*) Stakeholder Dialogues on Sustainable Development Strategies: Lessons, Opportunities and Developing Country Case Studies,* Environmental Planning Issues No 26, International Institute for Environment and Development, London

David, P (1995) *Responsabilisation villageoise et transferts de compétence dans le cadre de la gestion des terroirs au PNGT,* document de travail, 2è version, Octobre

Degnbol, T (1996) 'The terroir approach to natural resource management: panacea or phantom? – the Malian experience'. Working Paper No 2/1996, International Development Studies, Roskilde University, Denmark

de Graaf, M (1996) 'How to do it? Tools and challenges for donors in the implementation of CDE initiatives'. In: *Proceedings of the OECD/DAC Workshop on Capacity Development in Environment, Rome, Italy, 4–6 December,* pp195–210

Delli Priscolli, J (1997) 'Participation and conflict management in natural resources decision-making'. In: B Solberg and S Miina (eds) *Conflict Management and Public Participation in Land Management,* European Forestry Institute Proceedings No 14, Helsinki, pp61–87

Denniston, D (1995) *Defending the Land with Maps,* PLA Notes No 22, p38, Sustainable Agriculture Programme, International Institute for Environment and Development, London

Dent, D L (1988) 'Guidelines for land use planning in developing countries'. *Soil Survey and Land Evaluation* **8**(2):67–76

Dent, D (1997) 'Policy for the management of natural resources'. *The Land* **1**(1):3–26

Dent, D L and L K P A Goonewardene (1993) *Resource Assessment and Land Use Planning in Sri Lanka: A Case Study,* Environmental Planning Issues No 4, International Institute for Environment and Development, London

Dent, D L and A Young (1981) *Soil Survey and Land Evaluation,* George Allen and Unwin, London

Dewar, D, A Todes and V Watson (1986) *Regional Development and Settlement Policy,* Allen and Unwin, London

DeWit, C T, H van Keulen, N G Seligman and I Spharim (1988) 'Applications of interactive multiple goal programming techniques for analysis and planning of regional agricultural development'. *Agricultural Systems* **26**:211–30

Diarra, M. (1998) *Le PGRN: Présentation et bilan*, communication à la quatriéme rencontre réégionale des projets et programmes de GRN en Afrique de l'Ouest, 12–16 Novembre, Niamey, PGRN, Mali

DLARR/DIP (2001) *People First: Zimbabwe's Land Reform Programme*, Ministry of Lands, Agriculture and Rural Resettlement in conjunction with the Department of Information and Publicity, Office of the President, Government Printer, Harare, June

DNEF (1993) *Rapport de synthése de la conference nationale sur la relecture des texts forestiers, 28–30 Juillet 1993, Bamako, Mali*, Direction National des Eaux et Forêts, Mali

Donnelly, A, D B Dalal-Clayton and R Hughes (1998) *A Directory of Impact Assessment Guidelines. Second Edition*, International Institute for Environment and Development, London

DPCSD (1997) *Indicators of sustainable Development: Framework and Methodologies*, Division for Sustainable Development, Department for Policy Coordination, Commission for Sustainable Development, United Nations, New York

Driessen, P M and Konijn, N T (1991) *Land Use Systems Analysis*, INRES, Wageningen Agricultural University, The Netherlands

Dubois, O (1994) 'Poverty and environment: lessons from some cases in the Philippines and Southeast Asia'. Paper prepared for the International Workshop on Poverty and Environment, Belem, Brasil, 8–10 June

Dubois, O (1997) 'Decentralisation and local management of forest resources in sub-Saharan Africa: let it go or let it be (laissez faire)? A comparative analysis'. Unpublished review for the Policy that Works Project, Forestry and Land Use programme, International Institute for Environment and Development, London

Dubois, O (1998a) *Capacity to Manage Role Changes in Forestry: Introducing the '4Rs' Framework*, IIED Forest Participation Series No 11

Dubois, O (1998b) 'Getting participation and power right in collaborative forest management: can certification and the '4Rs' help? Lessons from Africa and Europe'. Master's Thesis for the European Programme in Environmental Management

Dubois, O, B Gueye, R Moorhead and F Ouali (1996) 'Etude des points forts et faibles des projets forestiers financés par les Pays-Bas en Afrique occidentale sahélienne'. Etude préparée pour les Ambassades des Pays-Bas de Bamako, Ouagadougou et Dakar, Novembre

Duivenbooden, N van, P A Gosseye and H van Keulen (eds) (1991) 'Competing for limited resources: the case of the Fifth Region of Mali: Report 2, Plant, livestock and fish production'. Étude sur les Systèmes de Production Rurales (ESPR), Mopti, Mali. DLO Centre for Agrobiological Research (CABO-DLO) Wageningen

Dumanski, J and E Craswell (1996) 'Resource management domains for evaluation and management of agro-ecological systems'. In: J K Syers and J Bouma (eds) *International Workshop on Resource Management Domains*, IBSRAM Proc. 16, Bangkok, pp 3–13

Dunsmore, J R, A Blair-Raines, G D B Lowe, D J Moffat, I P Anderson and J B Williams (1976) *The Agricultural Development of The Gambia: An Agricultural, Environmental and Social Analysis*, Land Research Study 22, Land Resources Development Centre, Surbiton

Dykstra, D P (1984) *Mathematical Programming for Natural Resource Management*, McGraw-Hill series in forest resources, McGraw-Hill, New York

Dzumbe, R, M Mwasi, E Mweri, M Ngari, L Vijselaar and M Kitz (1995) *A Strategy Workshop Report for a Participatory and Integrated Development Approach (PIDA) in Kilifi District, Kenya*, 6–9 August, Kilifi Development Programme (KIWASAP), Kilifi Ditsrict, Kenya

ECDPM (1998) *Approaches and Methods for National Capacity Building*, Maastricht, ECDPM, September

Ellis, F (1998) 'Livelihood diversification and sustainable rural livelihoods'. In: D Carney (ed) *Sustainable Rural Livelihoods: What Contribution Can We Make?* DFID, London, pp53–66

Emery, M and F E Emery (1978) 'Searching'. In: J W Surherland (ed) *Management Handbook for Public Administrators*, Van Nosrand Reinhold, New York

Engel, A (1997) 'Decentralisation, local capacity and regional rural development: experiences from GTZ-supported initiatives in Africa'. Paper prepared for the Technical Consultation on Decentralisation, Food and Agriculture Organization of the United Nations, Rome, 16–18 December

Enserink, B (1998) *Improving Public Participation in Problem Exploration*, SEPA, mimeo, Delft University of Technology, The Netherlands

Esmail, T (1997) 'Designing and scaling-up productive natural resource management programs: decentralisation and institutions for collective action'. Paper prepared for the Technical Consultation on Decentralisation, Food and Agriculture Organization of the United Nations, Rome, 16–18 December

Evans, H E (1990) *Rural–urban Linkages and Structural Transformation*, Report INU 71, Infrastructure and Urban Development Department, The World Bank, Washington, DC

Evans, P (1996) 'Government action, social capital and development: reviewing the evidence on synergy'. *World Development* **24**(6):1119–32

Faguet, J-P (1997) 'Decentralisation and local government performance'. Paper prepared for the Technical Consultation on Decentralisation, Food and Agriculture Organization of the United Nations, Rome, 16–18 December

Fair, T S D (1990) *Rural-urban Balance: Policy and Population in Ten African Countries*, African Institute of South Africa, Pretoria

Fall, A S and A Ba (1997) 'Pauvreté en Milieu Rural: Quelles Stratégies de Lutte?' Paper prepared for the workshop *Interactions Villes-Villages au Sénégal: Un Etat des Lieux*, 27–9 October, Saly

FAO (1976) *A Framework for Land Evaluation*, Soils Bulletin 32, Food and Agriculture Organization of the United Nations, Rome

FAO (1978) *Report of the Agro-Ecological Zones Project, Vol. 1. Methodology and Results for Africa*, World Soil Resources Report 48, Food and Agriculture Organization of the United Nations, Rome

FAO (1984a) *Guidelines: Land Evaluation for Rainfed Agriculture*, Soils Bulletin 52, Food and Agriculture Organization of the United Nations, Rome

FAO (1984b) 'Land Evaluation for Forestry'. Forestry Paper 48, Food and Agriculture Organization of the United Nations, Rome

FAO (1985) *Guidelines: Land Evaluation for Irrigated Agriculture*, Soils Bulletin 55, Food and Agriculture Organization of the United Nations, Rome

FAO (1991) *Guidelines: Land Evaluation for Extensive Grazing*, Soils Bulletin 58, Food and Agriculture Organization of the United Nations, Rome

FAO (1993) *Guidelines for Land Use Planning*, Development Series 1, Food and Agriculture Organization of the United Nations, Rome

FAO (1995) *Understanding Farmers' Communication Networks: An Experience in the Philippines*. Communication for Development Case Study 14, Food and Agriculture Organization of the United Nations, Rome

FAO/UNDP (1969) *Mahaweli Ganga Irrigation and Hydropower Survey, Ceylon*, Final Report. Food and Agriculture Organization of the United Nations, Rome

FAO/UNEP (1996) *Our Land, Our Future: A New Approach to Land Use Planning and Management*. Food and Agriculture Organization of the United Nations, Rome

FAO/UNEP (1997) *Negotiating a Sustainable Future for Land*, Land and Water Development Division, Food and Agriculture Organization of the United Nations, Rome

Feeney P (1998) *Accountable Aid: Local Participation in Major Projects*, Oxfam Insights, Oxfam Publications

Fisher, R J (1995) *Collaborative Management of Forests for Conservation and Development: Issues in Forest Conservation*, The World Conservation Union (IUCN), Gland, Switzerland

Fizbein, A (1997) 'Decentralisation and local capacity: some thoughts on a controversial relationship'. Paper prepared for the Technical Consultation on Decentralisation, Food and Agriculture Organization of the United Nations, Rome, 16–18 December

Fox, J (1994) 'Latin America's emerging local politics'. *Journal of Democracy* April: 105–15

Franklin, B and D Morley (1992) 'Contextual searching: cases from waste management, nature tourism, and personal support'. In: M R Weisbord (ed) *Discovering Common Ground*, Berrett-Koehler, San Francisco, pp229–46

Gado, B A (1996) 'Une instance locale de gestion et de régulation de la compétition foncière: Rôle et limites des commssions foncières au Niger'. Communication présentée au Séminaire de Gorée sur l'Initiative Franco-Britannique relative au Foncier et la Gestion des Ressources Naturelles en Afrique de l'Ouest, 18–22 November

Gaile, G L(1992) 'Improving rural–urban linkages through small town market-based development'. *Third World Planning Review* **14**(2):131–48

Ghana MLGRD (1996) *Ghana: The New Local Government System*, Ministry of Local Government and Rural Development, Accra, Ghana

Ghana NDPC (1997) *Vision 2020: The First Medium-term Development Plan (1997–2000)*

Gill, G (1993) *OK, The Data's Lousy, But it's All We've Got (Being a Critique of Conventional Methods)*, Sustainable Agriculture Programme Gatekeeper Series SA38, International Institute for Environment and Development, London

Goldman, I (1998) 'Decentralisation and sustainable rural livelihoods'. In: D Carney (ed) *Sustainable Rural Livelihoods: What Contribution Can We Make?.*, Papers presented at the Department for International Development's Natural Resource Advisers' Conference, July. Department for International Development, London

Government of Zimbabwe (1994) *Report of the Commission of Inquiry into Appropriate Agricultural Land Tenure Systems* (chaired by Professor M. Rukuni), Government of Zimbabwe, Harare

Graziano da Silva, J (1995) 'Urbanização e pobreza no campo'. In: P Ramos and B Reydon (eds) *Agropecuária e Agroindústrie no Brasil Ajuste. Situação Atual e Perspectives* Edição ABRA

Grosjean, R, O Dubois, M D Diakité and M Leach (1998) *Evaluation interne du Programme Eco-Développement Participatif du FENU*, Rapport préparé pour le Fonds d'Equipement des Nations Unies (UN Capital Development Fund), February

GTZ (1984) *Regional Rural Development: Guiding Principles*, Deutsche Gesellschaft für Technische Zusammenarbeit (GTZ) GmbH, Eschborn, Germany

GTZ (1993) *Regional Rural Development RRD Update: Elements of a Strategy for Implementing the RRD Concept in a Changed Operational Context*, Deutsche Gesellschaft für Technische Zusammenarbeit (GTZ) GmbH, Eschborn, Germany

GTZ (1995) *Creating Local Agendas: (B) Participatory Planning and Evaluation Methods: Suggestions for Complementary Methodologies*, Abteilung 402 GTZ/GATE, Eschborn

GTZ (1996) *Process Monitoring: Work Document for Project Staff*, Doc. No 402/96 22e NARMS, Department 402

Guba, E G and Y Lincoln (1989) *Fourth Generation Evaluation*, Sage, London

Guijt, I (1991) *Perspectives on Participation: An Inventory of Institutions in Africa*, International Institute for Environment and Development, London

Guijt, I and Hinchcliffe, F (1998) *Participatory Valuation of Wild Resources: An Oveview of the Hidden Harvest Methodology*, Sustainable and Rural Livelihoods Programme, International Institute for Environment and Development

Hammond, A (1995) *A Systematic Approach to Measuring and Reporting on Environmental Policy Performance in the Context of Sustainable Development*, World Resources Institute, Washington, DC

Hansell, J R F and J R D Wall (1974–76) *Land Resources of the Solomon Islands*, 8 vols, Land Resource Study 18, Land Resources Division, Ministry of Overseas Development, Surbiton

Harding, D, J K Kiara and K Thomson (1996) 'Soil and water conservation in Kenya: the development of the catchment approach and structured participation led by the Soil and Water Conservation Branch of the Ministry of Agriculture, Kenya'. Paper prepared for the OECD/DAC Workshop on Capacity Development in the Environment, 4–6 December, Rome

Harrington, L H and R Tripp (1984) 'Recommendation domains: a framework for on-farm research'. Unpublished CIMMYT Economics Working Paper. CIMMYT, Mexico

Harriss-White, B (1995) 'Maps and landscapes of grain markets in South Asia'. In: J Harriss, J Hunter and C M Lewis (eds) *The New Institutional Economics and Third World Development*, Routledge, London, pp87–108

Hazell, P B R (1986) *Mathematical Programming for Economic Analysis in Agriculture*, Macmillan, New York

Hilhorst, T and A Coulibaly (1997) 'What opportunities do local conventions offer for a sustainable management of sylvo-pastoral areas? First experiences with the implementation of the new forest code in southern-Mali'. Paper presented at the conference on 'Implementation of Environmental policies in Africa, Amsterdam, 17–18 April

Hilhorst, T and A Coulibaly (1999) 'Elaborating a local convention for managing village woodlands in southern Mali'. Issue Paper No 78, Drylands Programme, International Institute for Environment and Development, London

Hill, I D (ed) (1978/79) *Land Resources of Central Nigeria. Agricultural Development Possibilities*, 7 vols. Land Resource Study 29, Land Resource Development Centre, Surbiton, England

Hills, R C (1981) 'Surveys, surveyors and development'. *Soil Survey and Land Evaluation* **1**(3):40–3

Hobley, M (1995) *Institutional Change within the Forest Sector – Centralised Decentralisation*, Rural Development Forestry Network, Overseas Development Institute, London

Hockensmith, R D and J G Steele (1949) 'Recent trends in the use of land capability classification'. *Proceedings of the Soil Science Society of America* **14**:383–8

Huber, M and C J Opondo (eds) (1995) *Land Use Change Scenarios for Subdivided Ranches in Laikipia District*, Laikipia Research Program (LRP) Rep 19, Kenya

Hughes, R, S Adnan and D B Dalal-Clayton (1994) *Floodplains or flood plans?: A critical look at approaches to water management in Bangladesh*, International Institute for Environment and Development, London, and Research and Advisory Services, Dhaka

Hunting Survey Corporation (1962) *A Report on a Survey of Resources of Mahaweli Ganga Basin, Ceylon*, HSC, Toronto, and Government Press, Colombo

Hunting Technical Services (1974) *Southern Darfur Land Use Planning Survey*, Borehamwood

Hunting Technical Services (1976) *Savanna Development Project, Phase II*, Borehamwood

Hunting Technical Services (1977) *Agricultural Development in the Jebel Marra Area*, Borehamwood

Hunting Technical Services (1979/80) *Victoria Scheme Mahaweli Development Project, System C Feasibility Study/Final Designs and Cost Estimate*, Borehamwood

Hurditch, B (1996) 'Changing roles in forest management and decision-making: an overview'. In: K L Harris (ed) *Making Forest Policy Work 1996: Conference Proceedings of the Oxford Summer Course Programme 1996*, Oxford Forestry Institute, pp67–73

Huybens, N (1994) *Cours de communication – Portefeuille de lectures*, Fondation Universitaire Luxembourgeoise, Arlon, Belgium

IIED (1994a) *Whose Eden? An Overview of Community Approaches to Wildlife Management*, International Institute for Environment and Development, London

IIED (1994b) *Economic Valuation of Tropical Forest Land Use Options: A Review of Methodology and Applications*, Environmental Economics Programme, International Institute for Environment and Development, London

IIED (1995) *The Hidden Harvest: The Value of Wild Resources in Agricultural Systems – A Summary*, Sustainable Agriculture Programme/Environmental Economics Programme, International Institute for Environment and Development, London

IIED (1998) *Participatory Valuation of Wild Resources: An Overview of the Hidden Harvest Methodology*, International Institute for Environment and Development, London

IIED/WCMC (1996) *Forest Resource Accounting: Stocktaking for Sustainable Forest Management*, International Institute for Environment and Development, London, and World Conservation Monitoring Centre, Cambridge

ITTO (1993) *Guidelines for the Establishment and Sustainable Management of Planted Tropical Forests*, Policy Development Series No 4, International Tropical Timber Organization, Yokohama

IUCN (1982) *Categories, Objectives and Criteria for Protected Areas*, International Union for Conservation of Nature and Natural Resources, Gland, Switzerland

IUCN-Vietnam (1998) 'A study on aid and the environment sector: lessons learned discussion paper'. Paper prepared for a donors' meeting, December (unpublished draft)

IUCN/WWF/UNEP (1980) *The World Conservation Strategy*, International Union for Conseration of Nature and Natural Resources, Gland, Switzerland

Ive, J R and K D Cocks (1988) 'LUPIS: A decision-support system for land planners and managers'. In: P W Newton, M A P Taylor and R Sharp *Desktop Planning: Microcomputer Applications for Infrastructure and Services Planning and Management*, Hargreen, Melbourne, pp129–39

Ive, J R, J R Davis and K D Cocks (1985) 'A computer package to support inventory, evaluation and allocation of land resources'. *Soil Survey and Land Evaluation* 5:77–87

Johansson, L and A Hoben (1992) 'RRAs for land policy formulation in Tanzania'. *Forests, Trees and People Newsletter 15/16*, Swedish Universities Agricultural Society, Uppsala

Kaimowitz, D, G Flores, J Johnson, P Pacheco, I Pavéz, M Roper, C Vallejos and R Vélez (1998) *Local Government and Biodiversity Conservation in the Bolivian Tropics*, Center for International Forestry Research, Bogor, Indonesia

Kamete, A Y (1998) 'Interlocking livelihoods: farm and small town in Zimbabwe'. *Environment and Urbanization* 10(1):23–34

Karsenty, A (1996) 'Entrer par l'outil, la loi, ou les consensus locaux?' In: P Lavigne Delville (ed) *Foncier Rural, Ressources Renouvelables et Développement: Analyse Comparative des Approches*, GRET, pp35–40

Kassam, A H, H T van Velthuizen, G W Fischer and M M Shah (1991) *Agroecological Land Assessment for Agricultural Development Planning. A Case Study in Kenya. Main Report and 8 Appendices*, Food and Agriculture Organization of the United Nations and International Institute for Applied Systems Analysis, Rome

Kasumba, G (1997) 'Decentralising aid and its management in Uganda: lessons for capacity-building at local level'. ECDPM Working Paper No 20

Kauzeni, A S, I S Kikula, S A Mohamed, J G Lyimo and D B Dalal-Clayton (1993) *Land Use Planning and Resource Assessment in Tanzania: A Case Study*, Environmental Planning Issues No 3, International Institute for Environment and Development, London

Kenting Earth Sciences Ltd (1986) *Land Resource Mapping Project. Land Systems Report*, Government of Nepal/Government of Canada, 140 + 53pp, one of 13 volumes

Kerr, J (1994) 'How Subsidies Distort Incentives and Undermine Watershed Development Projects in India'. Paper for IIED/ActionAid Conference New Horizons: The Social, Economic and Environmental Impacts of Participatory Watershed Development, Bangalore, India: 28 November–2 December

Kessell, S R (1990) 'An Australian geographical information and modelling system for natural area management'. *International Journal of Geographic Information Systems* 4:333–62

Khanya-mrc (2000) *Rural Planning in South Africa: A Case Study. A Report Prepared by Khanya–Managing Rural Change, Bloemfontein*, Environmental Planning Issues No 22, International Institute for Environment and Development, London

Khanya-mrc (2001) *Review of Community Based Planning in South Africa*, Khanya–Managing Rural Change, Bloemfontein

Kiara, J K, L S Munyikombo, L S Mwarasomba, J Pretty and J Thompson (1999) 'Impacts of the catchment approach to soil and water conservation: experiences of the Ministry of Agriculture, Kenya'. In: F Hinchcliffe, J Thompson, J Pretty, I Guijt and P Shah (eds) *Fertile Ground: The Impacts of Participatory Watershed Management*, IT Publications, London, pp130–142

Kievelitz, U (1995) *Rapid District Appraisal: An Application of RRA on District Planning Level*, Kurzinfo Nr 21, 8–12. Sektorúnberg Städtische und Ländliche Programme. GTZ, Eschborn

Kikula, I S, D B Dalal-Clayton, C Comoro and H Kiwasila (1999) 'A framework for district participatory planning in Tanzania, Vol 1. Report prepared by the Institute of Resource Assessment of the University of Dar es Salaam and the International Institute for Environment and Development, London', Ministry of Regional Administration and Local Government and the United Nations Development Programme, Tanzania

Kleemeier, L (1988) 'Integrated rural development in Tanzania'. *Public Administration and Development* **8**(1):61–74

Klein, G and A Mabin (1998) *Bringing Needs and Solutions Together by Making Communities Part of the Larger Whole in Development Planning*, National Theme Study: District Planning, National Land Reform Policy Programme, Land and Agriculture Policy Centre, Johannesburg

Klingebiel, A A and P H Montgomery (1961) *Land Capability Classification*, US Department of Agriculture Handbook No 210, Washington, DC

KMPND (1989) *Samburu District Development Plan: 1989–1993*, Ministry of Planning and National Development, Nairobi, Kenya

Kuik, O and H Verbruggen (1991) *In Search of Indicators of Sustainable Development*, Kluwer Academic, Dordecht

Kullenberg, L, R Shotton and L Romeo (1997) 'Supporting rural local governments in practice: issues for consideration'. Paper prepared for the Technical Consultation on Decentralisation, Food and Agriculture Organization of the United Nations, Rome, 16–18 December

Land Resources Department/Bina Program (1990) *The Land Resources of Indonesia: A National Overview* from *Regional Physical Planning Progam for Transmigration*. Land Resources Dept., Natural Resources Institute, Overseas Development Administration, London and Direktorat Bina Program, Jakarta. See also: *The Land Resources of Indonesia: A National Overview – Atlas*. Land Resources Dept., Natural Resources Institute, Overseas Development Administration, London, and Directorate General of Settlement Preparation, Jakarta

Lane, C and I Scoones (1991) 'Barabaig natural resource management'. Paper presented to Conference on the World Savannas: Economic Driving Forces, Ecological Constraints and Policy Options for Sustainable Land Use. 23–26 January, Unesco, Nairobi

Laue, R and A S Kruger (1995) *Institutionalisation of Participation: The Case of SARDEP, Namibia*, Kurzinfo No 21, April 1995, pp 26–31, Sektorúnberg Städtische und Ländliche Programme, GTZ, Germany

Le Roy, E (1996) 'Les orientations des réformes foncières depuis le début des années quatre-vingt dix'. In: P Lavigne Delville (ed) *Foncier Rural, Ressources Renouvelables et Développement: Analyse Comparative des Approaches*, GRET, pp249–54

Lincoln, Y S and E G Guba (1985) *Naturalistic Enquiry*, Sage, Newbury Park

Livingstone, I (1997) 'Rural industries in Africa: hope and hype'. In: D F Bryceson and V Jamal (eds) *Farewell to Farms: De-Agrarianisation and Employment in Africa*, Africa Studies Centre Leiden, Research Series 1997/10, Ashgate, Aldershot, pp205–21

Lockie, S and F Vanclay (eds) (1997) *Critical Landcare*, Key Papers Series No 5. Centre for Rural Social Research, Charles Sturt University, Wagga Wagga, NSW, Australia

Lund, C. (1996) 'Land tenure disputes and state community and local law in Burkina Faso'. SEREIN Working Paper 14

Lutz, E and J Caldecott (eds) (1996) *Decentralisation and Biodiversity Conservation*, A World Bank Symposium

Lutz, W (ed) (1996) *The Future Population of the World, Second Edition*, International Institute for Applied Systems Analysis/Earthscan Publications, London

Maatman, A H, C Sawadogo, C Schweigman and A Ouedraogo (1998) 'Application of zaï and rock bunds in the north-west region of Burkina Faso: a study of its impact on household level using a stochastic linear programming model'. *Netherlands Journal of Agricultural Science* **116**:123–36

Makano, R, R Sichinga and L Simwanda (1997) *Understanding Stakeholders' Responsibilities, Relationships, Rights and Returns in Forest Resource Utilisation in Zambia: What Changes are Required to Achieve Sustainable Forest Management?* Study for IIED, September 1997

Makin, M J, T J Kingham, A E Waddams, C J Birchall and T Tamene (1975) *Development Prospects of the Southern Rift Valley, Ethiopia*, Land Research Study 21, Land Research Division, Ministry of Overseas Development, Surbiton

Manor, J (1997) *The Political Economy of Decentralisation*, Draft Report for the World Bank, August 1997

Mansfield, J E, J G Bennett, R B King, D M Lang and R M Lawton (1975/76) *Land Resources of the Northern and Luapula Provinces, Zambia: A Reconnaissance Assessment*, 6 vols, Land Resource Study 19, Land Resources Development Centre, Surbiton

Marshall, C (1990) 'Goodness criteria: are they objective or judgement calls?' In: E G Guba (ed) *The Paradigm Dialogi*, Sage, Newbury Park

Martin, P and J Woodhill (1995) 'Landcare in the balance: Government roles and policy issues in sustaining rural environment'. *Australian Journal of Environmental Management* **2**:3

Mathieu, P (1996) 'Réformes législatives et pratiques foncières en situations de transition. Comment sécuriser?' In: *10 ans de réorganisation agraire au Burkina. La sécurisation foncière en question*, Faculté de Droit de l'Université de Ouagadougou

Matove, R, Y Sasya, J Masanje and M Makiya (1997) 'HIMA natural resources conservation and land use management projects, Tanzania'. In: S Power, P Makungu and A Qaraeen (eds) *Watershed Management, Participatory and Sustainable Development*. Proceedings of Danida's Second International Workshop on Watershed Development, Iringa Region, Tanzania, 25 May–5 June, pp151–168

Mayers, J and N A Kotey (1996) *Local Institutions and Adaptive Management in Ghana*, Forestry and Land Use Series No 7, International Institute for Environment and Development, London

Mayers, J and B Peutalo (1995) *NGOs in the Forest: Participation of NGOs in National Forestry Action Programmes: New Experience in Papua New Guinea*, Forestry and Land Use Series No 8, International Institute for Environment and Development, London

McCormack, D E and M A Stocking (1986) 'Soil potential rating I. An alternative form of land evaluation'. *Soil Survey and Land Evaluation* **6**(2):37–42

McElwee, P D (1994) 'Common property and commercialisation: Developing appropriate tools for analysis'. MSc dissertation, Oxford Forestry Institute

Meagher, K and A R Mustapha (1997) 'Not by farming alone'. In: D F Bryceson and V Jamal (eds) *Farewell to Farms: De-Agrarianisation and Employment in Africa*, Africa Studies Centre Leiden, Research Series 1997/10, Ashgate, Aldershot, pp63–84

Mellors, D R (1988) 'Serenje, Mpika and Chinsali districts integrated rural development program, Zambia'. Case Study 27 in: C Conroy and N Litvinoff (eds) *The Greening of Aid: Sustainable Livelihoods in Practice*, Earthscan Publications, London, pp227–34

MET (1994) *Land Use Planning: Towards Sustainable Development. Policy Document*, Ministry of Environment and Tourism, Windhoek

Mitchell, A J B (ed) (1984) *Land Evaluation and Land Use Planning in Tabora Region, Tanzania*, Land Resource Study 35, Land Resources Development Centre, Surbiton

Mosse, D (1995) *Social Analysis in Participatory Rural Development*, PLA Notes No 24, Sustainable Agriculture Programme, International Institute for Environment and Development, London

Mourik, D van (1987) 'Marketing and environmentally-oriented socio-economic quantitative land evaluation in Sejenane, Tunisia'. In: K J Beek et al (eds) *Quantified Land Evaluation Procedures*. ITC pub 6, Enschede, pp40–143

Muchena, F N and J van der Bliek (1997) 'Planning sustainable land management: finding a balance between user needs and possibilities'. *ITC Journal* **b**(4):229–34

Munemo, M (1998) 'The Zimbabwean District Environmental Action Plan (DEAP) as a national strategy for sustainable development'. Paper presented at the donor-developing country scoping workshop on national strategies for sustainable development, 18–19 November , Sunningdale, UK

Murdoch, G, J Ojo-Atere, G Colborne, E I Olomu and E M Odugbesan (1976) *Soils of the Western States of Savanna in Nigeria*, Land Resource Study 23, Land Resource Development Centre, Surbiton

Murdoch, G, R Webster and C J Lawrence (1971) *A Land System Atlas of Swaziland*, MVVE/Director of Military Survey, Christchurch. England

Murphree, M W (1995) *The Lesson from Mayenye: Rural Poverty, Democracy and Wildlife Conservation*, Wildlife and Development Series No 1, International Institute for Environment and Development, London

Mwalyosi, R and R Hughes (1998) *The Performance of EIA in Tanzania*, Environmental Planning Issues No 14, International Institute for Environment and Development, London

Narayan D. (1993) *Focus on Participation: Evidence from 121 Rural Water Supply Projects*, UNDP-World Bank Water Supply and Sanitation Program, World Bank, Washington, DC

NDPC (1995) *Planning Guidelines for the Preparation of Sectoral and District Development Plans*. National Development Planning Commission, Accra, Ghana, December

NRI/UST (1997) *Kumasi Natural Resources Management Research Project: Inception Report. Volume 1: Main Report*, Natural Resources Institute, University of Greenwich, and University of Science and Technology, Kumasi, Ghana

ODA (1995a) *Note on Enhancing Stakeholder Participation in Aid Activities*, Social Development Department, Overseas Development Administration, London

ODA (1995b) *Guidance Note on How to do Stakeholder Analysis of Aid Projects and Programs*, Social Development Division, Overseas Development Administration, London

ODA (1995c) *Guidance Note on Indicators for Measuring and Assessing Primary Stakeholder Participation*, Social Development Department, Overseas Development Administration, London

OECD/UNDP (2002) *Sustainable Development Strategies: A Resource Book*, Organisation for Economic Co-operation and Development/United Nations Development Programme/Earthscan Publications, London

OECD-DAC (1997a) *Shaping the 21st Century*, Organisation for Economic Co-operation and Development – Development Assistance Committee, Paris

OECD-DAC (1997b) *Principles in Practice: Capacity Development in Environment*, Organisation for Economic Co-operation and Development – Development Assistance Committee, Paris

OECD-DAC (1999) *Assisting developing countries with the formulation and implementation of national strategies for sustainable development: the need to clarify DAC targets and strategies*, Document No DCD/DAC(99)11, 14 April, endorsed by the High Level Meeting of the Development Assistance Committee of the Organisation for Economic Co-operation and Development, Paris, 11–12 May

OECD-DAC (2001) *The DAC Guidelines: Strategies for Sustainable Development: Guidance for Development Cooperation*, Organisation for Economic Co-operation and Development – Development Assistance Committee, Paris

Ollier, C D, C J Lawrance, R Webster and P H T Beckett (1969) *Land Systems of Uganda: Terrain Classification and Data Storage*, Military Engineering Experimental Establishment Report No 959. Christchurch, England

Olowu, D (1990) 'The failure of current decentralisation programmes in Africa'. In: J S Wunsh and D Olowu (eds) *Failure of the Centralised State: Institutions and Self-governance in Africa*, Westview, Boulder, CO, pp74–94

Ostrom, E (1990) *Governing the commons: the evolution of institutions for collective action*, Cambridge University Press, Cambridge

Painter T (1993) *Getting it Right: Linking Concepts of Action for Improving the Use of Natural Resources in Sahelian West Africa*, Issue Paper No 40, Drylands Programme, International Institute for Environment and Development, London

Painter, T, J Sumberg and T Price (1994) 'Your terroir and my "action space": implications of differentiation, mobility and diversification for the approche terroir in Sahelian West Africa'. *Africa* **64**:4

Pallot, J K (1999) 'Strategic forest information for policy and institutional development: building on forest resource accounting'. Paper prepared for the UK Department for International Development, International Institute for Environment and Development, London

Pearce, D (ed) (1991) *Blueprint 2: Greening the World Economy*, Earthscan Publications, London

PlanAfric (1997) *Operationalising the Community Action Project: Assessment Studies, volume 2 (case studies)*, Report prepared by PlanAfric for the Ministry of Public Service, Labour and Social Welfare and the World Bank, Bulawayo, Zimbabwe

PlanAfric (2000) *Rural Planning in Zimbabwe: A Case Study. A Report Prepared by PlanAfric, Bulawayo*, Environmental Planning Issues No 23, International Institute for Environment and Development, London

PlanAfric (2001) *Organisation Review of the Give a Dam Campaign, Zimbabwe*, PlanAfric, Bulawayo, Zimbabwe

Potts, D (1995) 'Shall we go home? Increasing urban poverty in African cities and migration processes'. *The Geographic Journal* **161**(3):245–64

Pretty, J N (1994) 'Alternative systems of inquiry for sustainable agriculture'. *IDS Bulletin* **25**(2):37–48, Institute of Development Studies, University of Sussex

Pretty, J N (1995) *Regenerating Agriculture: Policies and Practice for Sustainability and Self-reliance*. Earthscan Publications, London

Pretty, J N (1997) 'The sustainable intensification of agriculture: making the most of the land'. *The Land* **1**(1):45–64

Pretty, J N and P Shah (1994) *Soil and Water Conservation in the 20th Century: A History of Coercion and Control*, Rural History Centre Research Series No 1. University of Reading

Pretty, J N, I Guijt, J Thompson and I Scoones (1995a) *A Trainer's Guide for Participatory Learning and Action*, International Institute for Environment and Development, London

Pretty, J N, J K Kiara and J Thompson (1995b) 'Agricultural regeneration in Kenya: the catchment approach to soil and water conservation'. *Ambio* **24**(1):7–15

Price Gittinger, J (1982) *Economic Analysis of Agricultural Projects, Second edition*, Johns Hopkins University Press, Baltimore MD

Pruitt, D and J Z Rubin (1986) *Social Conflict*, Random House, New York

Querner, E and R A Feddes (1989) *Calculations of Agricultural Crop Production and Cost Benefit on a Regional Level*, Winand Staring Centre Rep. 13. Wageningen, The Netherlands

Rahnema, M (1992) 'Participation'. In: W Sachs (ed) *The Development Dictionary*, Zed Books Ltd, London

Rakodi, C (1990) 'Policies and preoccupations in rural and regional development planning in Tanzania, Zambia and Zimbabwe'. In: D Simon (ed) *Third World Regional Development: A Reappraisal*, Paul Chapman Publishing, London

Reij, C (1988) 'The agroforestry project in Burkina Faso: An analysis of popular participation in soil and water conservation'. In: C Conroy and M Litvinoff (eds) *The Greening of Aid*, Earthscan Publications, London, pp74–7

Ribot, J C (1998a) 'Theorising access: forest profits along Senegal's charcoal commodity chain'. *World Development* **29**(2):307–341

Ribot, J C (1998b) *Decentralisation, Participation and Accountability in Sahelian Forestry: Legal Instruments of Political-administrative Control*, Center for Population and Development Studies, Harvard University

Ribot, J C (1999) *Integral Local Development: Authority, Accountability and Entrustment in Natural Resource Management*, Regional Program for the Traditional Energy Sector (RPTES), World Bank, Washington, DC

Richards, M, G Navarro, A Vargas and J Davies (1996) *Decentralising Forest Management and Conservation in Central America*, ODI Working Paper No 93, June

Rikken, G (1993) *Natural Resources Management by Self-help Promotion in the Philippines*, Asian Social Institute, Manila

Robertson, V C and R F Stoner (1970) 'Land use surveying: A case for reducing the costs'. In: Boesch et al (eds) *New Possibilities and Techniques for Land Use and Related Surveys*, Geographical Publications Ltd., London, pp3–15

Romeo, L (1997) *Issues in Analysis and Design of Local Development Funds*, UNCDF, New York

Romeo, L (1998) *Decentralised Development Planning: Issues and Early Lessons from UNCDF-supported 'Local Development Funds' (LDF) Programmes*, UNCDF, New York

Rondinelli, D and K Ruddle (1978) *Urbanization and Rural Development: A Spatial Policy for Equitable Growth*, Praeger, New York

Rossiter, D G and A R van Wambeke (1993) *ALES Version 4 User's Manual*, SCAC Teaching Series T93-2, Soil, Crop and Atmospheric Sciences Dept., Cornell University, Ithaca, NY

Sazanami, H and R Newels (1990) 'Subnational development and planning in Pacific island countries'. In: D Simon (ed) *Third World Regional Development: A Reappraisal*, Paul Chapman Publishing, London

Schubert, B, A Addai, S Kachelriess, J Kienzle, M Kitz, E Mausolf and H Schadlich (1994) *Facilitating the Introduction of a Participatory and Integrated Development Approach (PIDA) in Kilifi District, Kenya. Vol. 1: Recommendations for institutionalising PIDA based on four pilot projects*, Landwirtschaftlich-Gärtnerische Fakultät, Humboldt-Universität zu Berlin, Margraf Verlag, Weikersheim, Germany

Scoones, I (1998) *Sustainable Rural Livelihoods: A Framework for Analysis*, Working Paper No 72, Institute of Development Studies, University of Sussex

Scott, R M, R Webster and C J Lawrance (1971) *A Land System Atlas of Western Kenya, with Map 1:500 000*, Military Engineering Experimental Establishment, Christchurch, England

Seppala, P (1996) 'The politics of economic diversification: reconceptualizing the rural informal sector in Southeast Tanzania'. *Development and Change* **27**:557–78

Serageldin, I and A Steer (1994) 'Epilogue: expanding the capital stocks'. In: I Serageldin and A Steer *Making Sustainable Development Sustainable: From Concept to Action*, Environmentally Sustainable Development Occasional Paper Series No 2, World Bank, Washington, DC

Shah, M K (1995) 'Participatory reforestation experience from Bharuch District, South Gujarat, India'. *Forests, Trees and People Newsletter* No 26/27, April

Shaxson, T F, N D Hunter, T R Jackson and J R Alder (1977) *A Land Husbandry Manual*, Ministry of Agriculture, Lilongwe, Malawi

Sidaway, R (1997) 'Outdoor recreation and conservation: conflicts and their resolution'. In: B Solberg and S Miina (eds) *Conflict Management and Public Participation in Land Management*, European Forestry Institute Proceedings No 14, pp289–301

Siffin, W J (1980) *The Art of Problem Defining*, Pasitan, Indiana University, Bloomington IN

Simon, D (ed) (1990) *Third World Regional Development: A Reappraisal*, Paul Chapman Publishing, London

Simon, D and Rakodi, C (1990) 'Conclusions and prospects: what future for regional planning?' In: D Simon (ed) *Third World Regional Development: A Reappraisal*, Paul Chapman Publishing, London

Sinha, J K (1998) 'Improving the quality of peoples' participation in development projects'. *Exchanges* **22**, September, ActionAid

Smit, B, M J Brklacich, J Dumanski, K B MacDonald and M Miller (1984) 'Integral land evaluation and its application to policy', *Canadian Journal of Soil Science* **64**:467–79

Smith, J K (1990) 'Alternative research paradigm and the problems of criteria'. In: E Guba (ed) *The Paradigm Dialog*, Sage, Newbury Park

Smith, L D (1997) 'Decentralisation and rural development: the role of the public and private sectors in the provision of support services'. Paper prepared for the Technical Consultation on Decentralisation, Food and Agriculture Organization of the United Nations, Rome, 16–18 December

Smoke, P and Romeo, L (1997) 'Designing intergovernmental fiscal relations and international finance institutions allocations for rural development'. Paper prepared for the Technical Consultation on Decentralisation, Food and Agriculture Organization of the United Nations, Rome, 16–18 December

Soil Survey Staff (1951) *Soil Survey Manual*, USDA Handbook 436, Soil Conservation Service, Washington, DC

Stocking, M A and D E McCormack (1986) 'Soil potential ratings II. A test of the method in Zimbabwe'. *Soil Survey and Land Evaluation* **6**(3):115–22

Storie, R E (1933) 'An index for rating the agricultural value of soils'. *Bulletin of California Agricultural Experimental Station* **536**

Storie, R E (1978) *The Storie Index Soil Rating Revised*, Special Publication of the Division of Agricultural Sciences, University of California, No 3203

Sullivan, S (2001) 'Rural planning in Namibia: state-led initiatives and some rural realities'. In: D B Dalal-Clayton, D Dent and O Dubois (2000) *Rural Planning in the Developing World, with a Special Focus on Natural Resources: Lessons Learned and Potential Contributions to Sustainable Livelihoods. An Overview*, Environmental Planning Issues No 20, International Institute for Environment and Development, London

Sys, C, E Van Ranst and J Debaveye (1991) *Land Evaluation: Part I Principles in Land Evaluation and Crop Production Calculations, Part II Methods in Land Evaluation*, ITC University Ghent, Belgium

Sys, C, E Van Ranst, J Debaveye and F Beernaert (1993) *Land Evaluation, Part III: Crop Requirements*, Agricultural Publications 7, General Administration for Development Cooperation, Brussels

Tengberg, A, M A Stocking and M Da Viega (1997) 'The impact of erosion on the productivity of a Ferralsol and a Cambisol in Santa Catarina, Brazil'. *Soil Use and Management* **13**(2):906

Tendler, J and S Freedheim (1994) 'Trust in a rent-seeking world: health and government transformed in south east Brazil'. *World Development* **22**(12):1771–991

Thirion, S (1997) 'Local partnership as a key tool for participation and decentralisation'. Paper prepared for the Technical Consultation on Decentralisation, Food and Agriculture Organization of the United Nations, Rome, 16–18 December

Thomas, G, J Anderson, D Chandrasekharan, Y Kakabadse and V Matiru (1996) 'Levelling the playing-field: promoting authentic and equitable dialogue under inequitable conditions'. In: Volume 1: *Compilation of discussion papers made to the electronic conference on 'Addressing natural resource conflicts through community forestry'*, Food and Agriculture Organization of the United Nations, Rome, pp165–80

Thomas, P, J A Varley and J E Robinson (1979) *The Sulphidic and Associated Soils of the Gambia Estuary above the Proposed Barrage at Yelitenda, The Gambia*. Project Rep 89. LRDC, Surbiton

Thomas, S (1995) *The Legacy of Dualism in Decision-making within CAMPFIRE*, Wildlife and Development Series No 4, International Institute for Environment and Development, London

Thomson, J T and C Coulibally (1994) *Decentralisation in the Sahel: regional synthesis*. Regional Conference on Land Tenure and Decentralisation in the Sahel, Praia (Cape Verde), CILSS – Club du Sahel

Toulmin, C (1993) 'Gestion de terroir. Principles, first lessons and implications for action'. Discussion paper prepared for UNSO, Drylands Programme, International Institute for Environment and Development, London

Triantafilis, J and A B McBratney (1993) *Applications of Continuous Methods of Soil Classification and Land Suitability Assessment in the Lower Namoi Valley*, Divisional Report 121, CSIRO Division of Soils, Canberra

Turner, R K (1985) 'Land evaluation: financial, economic and ecological approaches'. *Soil Survey and Land Evaluation* 5(2):21–34

Turner, R K, D W Pearce and I J Bateman (1994) *Environmental Economics: An Elementary Introduction*, Harvester Wheatsheaf, Hemel Hempstead

UNCED (1992) *Agenda 21*, United Nations Conference on Environment and Development (UNCED), United Nations General Assembly, New York

UNCHS (1996) *An Urbanizing World: Global Report on Human Settlements 1996*, United Nations Centre for Human Settlements, Oxford University Press, Oxford

UN DESA (2002) *Report of an Expert Forum on National Strategies for Sustainable Development*, Meeting held in Accra, Ghana, on 7–9 November 2001. Department of Economic and Social Affairs, United Nations, New York (available at www.johnnesburgsummit.org)

UNDP/UNCHS (Habitat) (1995) 'Rural–urban linkages: Policy guidelines for rural development'. Paper prepared for the 23rd Meeting of the ACC Sub-Committee on Rural Development, UNESCO Headquarters, Paris, 31 May–2 June

UNGA (2001) *Report of the Secretary General: Road Map Towards the Implementation of the United Nations Millennium Declaration*, A/56/326, 6 September, United Nations, New York

United States Bureau of Reclamation (1953) *Bureau of Reclamation Manual, Vol V. Irrigated Land Use. Pt 2 Land Classification*, USBR, Denver, CO

UNSO (1994) *Gestion de terroirs: analyse de l'état d'avancement de la mise en oeuvre de l'approche dans la zone soudano-sahélienne de l'Afrique occidentale*. Compte-rendu de l'Atelier régional sur l'harmonisation et opérationallisation du concept de gestion de terroirs dans une perspective de développement durable, Niamey

van Keulen, W F and S J E Walraven (1996) 'Negotiations in participation: improving participatory methodologies with insights from negotiation theories'. Thesis Report, Department of Communication and Innovation Studies, Wageningen Agricultural University, August

Van Mourik (1987) see Mourik (van)

Veeneklas, F R, S Cissé, P A Gosseye, N van Duivenbooden and H van Keulen (1991) *Competing for Limited Resources: The Case of the Fifth Region of Mali. Report 4: Development scenarios*, CABO/ESPR, CABO-DLO, Wageningen

Vodoz, L (1994) 'La prise de décision par consensus: pourquoi, comment, à quelles conditions'. *Environnement & Société*, No 13, FUL, 55–66

Wade, R (1988) *Village Republics: Economic Conditions for Collective Action in South India*, Cambridge University Press, Cambridge

Walker, G B and S E Daniels (1996) 'Collaborative learning: improving public deliberation in ecosystem-based management'. *Elsevier Impact Assessment Review* **16**: 71–102

Walker, G B and S E Daniels (1997a) 'Foundations of natural resource conflict: conflict theory and public policy'. In: B Solberg and S Miina (eds) *Conflict Management and Public Participation in Land Management*, European Forestry Institute Proceedings No 14, Helsinki, pp13–33

Walker, G B and S E Daniels (1997b) 'Rethinking public participation in natural resource management: concepts from pluralism and five emerging approaches'. Paper prepared for the Food and Agriculture Organization of the United Nations Workshop on Sustainable Forestry and Rural Development in Pluralistic Environments, Rome, Italy, 9–12 December

Warren, D M (1988) 'A comparative assessment of Zambian integrated rural development programs'. *Manchester Papers on Development* **IV**(1):89–100

Watson, R T (1999) 'New strategies, strengthened partnerships'. *Environment Matters*, The World Bank, Washington, DC, pp 4–7

WCED (1987) *Our Common Future*, Report of the World Commission on Environment and Development, Oxford University Press, Oxford

WCFSD (1999) *Our Forest, Our Future*, Report of the World Commission on Forests and Sustainable Development, Cambridge University Press, Cambridge

Weber, J (1996) *Conservation, développement et coordination: peut-on gérer biologiquement le social?* Communication présentée au cours du colloque pan Africain sur la gestion communautaire des ressources naturelles renouvelables et développement durable, 23–27 June, Harare, Zimbabwe

Wheeler, J R, R J White and M R Wickstead (1989) *Land Use Planning Project, Tabora Region, Tanzania. Final report*, Overseas Development Natural Resources Institute, Chatham

Wilson, M J and T E Bourne (1971) *North Nyamphande Settlement Scheme*, Land Use Services Division, Ministry of Rural Development, Chipata, Zambia

Wilson, M J and J R Priestley (1974) *Msandile Catchment Plan*, Land Use Services Division, Ministry of Rural Development, Chipata, Zambia

Wily, L (1996) *Villagers as Forest Managers and Governments 'Learning to Let Go': The Case of Duru Haitemba and Ngori Forests in Tanzania*, Forest Participation Series No 9, International Institute for Environment and Development, London

Winckler, G, R Rochette, C Reij, C Toulmin, and E Toé (1995) *Approche gestion des terroirs au Sahel: analyse et évolution. Mission de dialogue avec les projets GT/GR du Club du Sahel au Burkina Faso, au Niger et au Mali, 1994–95. Rapport de mission*, Comité Permanent Inter-États de Lutte Contre la Sécheresse dans le Sahel (Cilss) et Club du Sahel, Ouagadougou, Burkina Faso

Winograd, M (undated) *Indicators for Latin America and the Caribbean: towards land use sustainability*, World Resources Institute, Washington, DC

Wolf, E R (1957) 'Closed corporate peasant communities in Mesoamerica and Central Java'. *Southwestern Journal of Anthropology* **13**

Woode, P R (1981) 'We don't want soil maps. Just give us land capability'. The role of land capability surveys in Zambia'. *Soil Survey and Land Evaluation* **1**:2–5

Woodhouse, P (1997) 'Governance and local environmental management in Africa'. *Review of African Political Economy* **74**:537–47

World Bank (1994) *Report of the Learning Group on Participatory Development*, Third Draft, The World Bank, Washington, DC

World Bank (1995) *Pakistan: The Aga Khan Rural Support Programme. A Third Evaluation*, OED, World Bank, Washington, DC

World Bank/IMF (1999) Communiqué of the Joint Ministerial Committee of the Boards of Governors of the World Bank and the International Monetary Fund on the Transfer of Real Resources to Developing Countries. Washington, DC

Yacouba, M (1996) *L'expérience nigérienne en matière de gestion des ressources naturelles: rôle des commissions foncières*, Communication présentée au Séminaire de Gorée sur l'Initiative Franco-Britannique relative au Foncier et la Gestion des Ressources Naturelles en Afrique de l'Ouest, 18–22 Novembre

Yacouba, M, C Reij and R Rochette (1995) *Atelier de restitution sur la gestion des terroirs et développement local au Sahel. Niamey, 30 Mai–2 Juin, compte-rendu des travaux*, Comit Permanent Inter-États de Lutte Contre la Sécheresse dans le Sahel (Cilss) and Club du Sahel, Ouagadougou, Burkina Faso

Yambo, A C, S Salmo and P Alino (1998) 'A preliminary assessment of coastal development planning in the municipal waters of Bolinao, Pangasinan, Philippines'. *Out of the Shell: Coastal Waters Network Newsletter* 6(2):April

Yan Zheng (1995) 'Township and village enterprise in China'. Fujian Academy of Sciences, mimeo

Young, A and P Goldsmith (1977) 'Soil survey and land evaluation in developing countries: a case study in Malawi'. *Geographical Journal* **143**:407–31

Young, R (1983) *Canadian Development Assistance to Tanzania*, The North–South Institute, Ottawa

Zambia Department of Agriculture (1977) *Land Use Planning Guide*, Department of Agriculture, Ministry of Lands and Agriculture, Lusaka

Zazueta, A (1995) *Policy Hits the Ground: Participation and Equity in Environmental Policy-making*, World Resources Institute, Washington, DC

Zoomers, A E B and J Kleinpenning (1996) 'Livelihood and urban–rural relations in Central Paraguay', *Tijdschrift voor Economische en Sociale Geografie* **87**(2):161–74

# Index